BIOLOGY THE EASY WAY

Gabrielle I. Edwards

Assistant Principal Supervision
Science Department
Franklin D. Roosevelt High School
Brooklyn, New York

BARRON'S

Barron's Educational Series, Inc.
Woodbury, New York / London / Toronto / Sydney

All inquiries should be addressed to:
Barron's Educational Series, Inc.
113 Crossways Park Drive
Woodbury, New York 11797

Library of Congress Catalog No. 84-343

International Standard Book No. 0-8120-2625-X

Library of Congress Cataloging in Publication Data
Edwards, Gabrielle I.
  Biology the easy way.

  1. Biology. i. Title.
QH307.2.E38 1984   574   84-343
ISBN 0-8120-2625-X

PRINTED IN THE UNITED STATES OF AMERICA

67 410 9876543

CONTENTS

PREFACE

BIOLOGY THE EASY WAY presents an up-to-date account of the principles and concepts of modern biology. Over the past 25 years there has been an explosion of knowledge in the biological sciences, which has revolutionized our way of thinking and doing in terms of biology. As a consequence of the many new discoveries in biology, some of the traditional beliefs have been discarded, others have been modified and much new material has been added to replace the invalid theories of old. A direct result of the increase in biological knowledge has been the increase in size and weight of the modern biology textbook. Although all of the newer knowledge in biology cannot be contained under one cover, the standard biology text has become a volume of substantial weight and cost.

BIOLOGY THE EASY WAY was prepared with the reader in mind. Although it contains the important information of modern biology, the writing is concise. The content material is presented clearly in language that is easy to read and to understand. The clarity of language serves to make difficult concepts more easily understood. Technical language is avoided wherever possible. However, biology has its special vocabulary which is essential for description. The words new to the reader are presented in italics, clearly defined and used appropriately in the text material. The overall quality of the writing in BIOLOGY THE EASY WAY is lively, modern and interesting.

BIOLOGY THE EASY WAY is divided into 16 chapters. Each of these chapters represents an area of specialization in the field of biology. For each such area modern principles of biology are presented in an appealing way for the reader. In each chapter, the content discussion is accompanied by carefully placed line drawings of the organisms, organs, structures or processes under discussion. Wherever possible, summary material has been presented in tabular form, providing a quick means of study and review of a given topic.

Each chapter in BIOLOGY THE EASY WAY provides special study aids that are designed to enhance the learning and understanding of the biological principles or concepts under study. Because scientific discovery and invention cannot be divorced from the history of human life on earth, a Chronology of Famous Names has been included in each chapter. Men and women scientists representing all countries, races, and religions are included in the chronology. The contributors mentioned range in time from the ancient Greek philosopher-scientists to modern day investigators. For each name the approximate date of the discovery or invention, the contributor's nationality, and a one line summary of the accomplishment is given. Another study aid included in this book is a listing titled Words For Study. The important words of biology introduced in the chapter are listed for review. Readers who take the time and effort to learn all of these words will measurably increase their knowledge of modern biology. At the end of each chapter are at least 30 questions which are designed to assess the reader's grasp of some important facts and concepts of biology. An answer key is provided to help with self-evaluation.

By reading the Table of Contents you can see the range of topics that is included in this text. By design, BIOLOGY THE EASY WAY is versatile, having appeal and use for students in secondary school through first year college and also to adults who are not in school but who desire to learn some modern biology. A chapter may be read in its entirety or in parts. The information is presented so that one can skim or read deeply. The index provides topic-page information.

The principle aim of BIOLOGY THE EASY WAY has been to present the facts of biology as they are known today in such a way that the curiosity and interest of the reader is aroused.

Gabrielle I. Edwards
January, 1984

CHAPTER 1

BIOLOGY: THE SCIENCE OF LIFE

BIOLOGY DEFINED

Biology is the science that studies life and living things, including the laws that govern the phenomena of life.

Every aspect of life from the smallest submicroscopic living particle to the largest and most imposing of plant and animals species is included in the study of biology. Biological study encompasses all that is known about any plant, animal, microbe or other living thing of the past or present.

Biology is a *natural* science because it is the study of organic (living) nature. It is the science of fishes and fireflies, grasses and grasshoppers, humans and mushrooms, flowers and starfish, worms and molds. It is the study of life on top of the highest mountain and at the bottom of the deepest sea. Biology is the accumulated knowledge about all living things and the principles and laws that govern life. Those who specialize in biology are known as *biologists* or *naturalists*, and it is through their observations of nature and natural phenomena that the great ideas of biology have been born.

BIRTH OF BIOLOGY

"Truth is the daughter of time" is an adage worth repeating. History has shown that only within a certain time frame of thought could the major ideas of biology have evolved. The birth of biology as a *bona fide* science was slow and painful, taking place over many centuries. The object of scientific study is to find the truth. So it was and still is that human beings have sought the truth about the nature of life.

The study of life is as old as humankind, dating back to ancient peoples who observed and wondered about the characteristics of the animals and plants around their limited sphere. The ancients used their observations to help them in such activities as hunting, food gathering and crop growing.

1

A great deal of credit is due to the ancient Greeks for having begun a systematic study of living things, including human beings. A brief chronology of ancient achievements follows:

Hippocrates (460–370 B.C.) founded the first medical school on the Greek island of Cos.

Aristotle (384–322 B.C.) founded the systematized study of natural history. He was a keen observer and a prolific writer-illustrator on subjects dealing with plants and animals.

Theophrastus (380–287 B.C.) founded the organized study of plants. He is called the "ancient father of botany."

Galen (A.D. 130–200) founded the science of anatomy. The accuracy of his anatomical drawings of animal bodies and the gross structure of the human body remained unchallenged for centuries.

The rise of ancient science reached a peak and then declined sharply. For several centuries interest and activity in scientific investigation waned, creating a void in scientific discovery. This depressed period in which there was little or no inquiry about nature and life is known historically as The Dark Ages; it lasted from A.D. 200–1200. During this time, books were scarce and authority was considered to be the source of all knowledge. Observation was considered to be impious prying.

The 14th century ushered in a revival of scientific thought and inquiry. Among the reasons for the change in attitude were the invention of the printing press, the voyages of the explorers, the expansion of ideas brought on by the Crusades and the rise of universities. All of these things contributed to a return to the study of nature and to the methods of science.

A thumbnail sketch of some important contributions to the understanding of the laws of natural science by early investigators follows:

Andreas Vesalius (1514–1564) refuted the authority of Galen and studied the human body by dissection.

Marcello Malpighi (1628–1694), an expert in plant and insect anatomy, described the metamorphosis of the silkworm.

Robert Hooke (1635–1703) discovered and named the "cells" in cork.

William Harvey (1578–1667) demonstrated the circulation of the blood in the human body.

Anton Von Leeuwenhoek (1632–1723) was the first person to see living cells.

Carolus Linnaeus (1707–1778) devised the system of binomial nomenclature by the double naming of species.

Georges Cuvier (1769–1832) founded the study of comparative anatomy.

Jean Baptiste Lamarck (1744–1829) coined the word *biology* by putting together two Greek words: *bios* meaning "life" and *logos* meaning "study."

Branches of Biology _____

Biology is an extensive science encompassing many life science disciplines that cover enormous areas of study and information. Estimations based on biostatistical methods indicate that the total number of living plant and animal species is probably about ten million. Scientists agree that not all living species have been discovered and perhaps only 15 percent of the total have been described. *Paleobiologists*, biologists who specialize in the study of ancient life, believe that the number of extinct species may range between 15 and 16 million.

Investigators of the 18th and 19th centuries were faced with the monumental task of trying to put some kind of order into the study of natural science. Thus, work during these centuries involved sorting, identifying, naming, describing and classifying the myriads of species that were being discovered. Biology was then divided into a few discrete areas which allowed for specialization of study.

Botany is the study of plants and plant life cycles. *Zoology* is the study of animals and their life histories. *Morphology* is the study of the structure and form of living things. There are two sub-areas of this discipline—namely, plant morphology and animal morphology. *Physiology* is that branch of biology that deals with the way living things function. It explains the way a whole organism and its parts work. The study of function may be related to plants or to animals. *Taxonomy* is the science of classification. It provides a logical and consistent method for naming and grouping the many species of living things.

TABLE 1.1. Branches of Botany

Botany is the systematized study of plants.	
Disciplines of Botany	*Study of*
bryology	mosses and liverworts
paleobotany	fossil plants
pteridology	ferns
palynology	modern and fossil pollen
plant pathology	plant diseases

TABLE 1.2. Branches of Zoology

Zoology is the systematized study of animals.	
Disciplines of Zoology	*Study of*
entomology	insects
ethology	animal behavior
herpetology	amphibians and reptiles
invertebrate zoology	animals without backbones
ichthyology	fishes
mammalogy	mammals
ornithology	birds
vertebrate zoology	animals with backbones

To accommodate the explosion of knowledge in the 20th century the number of biological science specialities increased. The modern study of life science takes into consideration levels of biological organization. Using the knowledges of chemistry, physics and mathematics, the modern biologist inquires into finite structure and function.

Molecular biology explains the roles of elements, compounds and ions in the fundamental biochemical processes of cells and their parts. *Energetics* is that branch of science used to explain the vital energy transformations that take place in living cells.

Above the molecular level, the major biological sciences present organized information about parts of organisms or about whole individuals or populations. *Morphology*, the study of structure, is now divided into three sub-fields: anatomy, histology and cytology. *Anatomy* is the study of gross structure that is visible to the naked eye. *Histology* is the study of microscopic structure on the tissue level of organization. *Cytology* is the study of cells on the microscopic level.

Taxonomy is a study of whole organisms. It is the science of classification by which living species are grouped according to observed similarities and presumed evolutionary relationships in structure and molecular organization. *Systematics* includes the identification and ranking of all plants, including their classification and nomenclature; it is the science that provides understanding about plant diversities.

Embryology is the science of embryo development in plants and animals. Embryology commences with the union of sperm and egg and continues to the completion of body structures.

Genetics is a specialty of biology that explains heredity and variation in living things. How molecules bearing genetic information are passed from parent to offspring and how they function in the cellular environment are important aspects of genetic studies.

Evolution is the science that determines the possible origins and relationships of living things and analyzes the changes that occur in species. Modern classification is based upon evolutionary relationships.

Paleontology is the study of ancient life accomplished through examination of fossil records.

Ecology is the study of relationships of plants and animals to one another and to their physical environment.

The Work of the Modern Biologist

It is through the painstaking efforts of biologists that so much information about living species of the present and past has been gathered. The methods used to gather facts are as varied as the disciplines of biological science and the myriad of scientists who work to unravel the secrets of life. Some basic principles direct the efforts and interests of those men and women who devote their lives to the demands of research in biology. Two of these principles are summarized below:

Homeostasis is the term used to describe the stable internal environment of the cell and the organism as a whole. Homeostasis is a condition necessary for life. The internal chemical balance of the cell must be maintained at a steady state to promote innumerable biochemical activities that foster the production and use of energy. The concept of homeostasis was initially developed by the 19th century physiologist Claude Bernard.

Unity is shared by all living species in that they have certain biological, chemical and physiological characteristics in common. There is unity of the basic living substance—*protoplasm*. Rudolph Virchow, a 19th century pathologist, established the concept of continuity of the germ plasm in which he stated, "All living cells arise from pre-existing living cells." All cells synthesize and use enzymes. Their genetic information is carried by DNA molecules. DNA gives cells the ability to replicate by a pattern which is followed in all cells.

THE SCIENTIFIC METHOD AND PROBLEM SOLVING

The methods which enable the biophysicist to measure particles too small to be resolved, or distinguished, by an electron microscope, the biologist to discover the cause of diseases such as toxic shock syndrome, the geologist to determine the age of the earth, and the biochemist to identify minute quantities of enzyme in a respiratory pathway are not the same as those used by the average person to solve the problems of daily living.

Scientific problem solving depends upon accuracy of observation and precision of method. Inherent in scientific thinking is orderliness of approach which invites the forming of conclusions from hypotheses, theories, principles, generalizations, concepts and laws. Scientific problem solving follows a pattern of behaviors which are collectively known as the *scientific method*.

The first step in the scientific method is *observation*. Observation is an accurate sighting of a particular occurrence. The accuracy of an observation becomes substantiated when a number of independent observers agree on seeing the same set of circumstances. For example: early in the 16th century, Vesalius presented the first accurate description of the arteries and veins in the human body. His drawings were based on first-hand observations obtained through dissection of human bodies. Over a period of 400 years succeeding generations of anatomists have attested to the accuracy of Vesalius' work. Moving away from our example, it is to be noted that although several trained observers may agree upon the same set of observations, the conclusions they draw from these observations may differ. Therefore, problem solving depends on other activities which enable the investigator to form conclusions based on accurate observations.

An observation becomes useful only in terms of the conclusion that can be drawn from it. A tentative conclusion derived from a set of observations is known as a *hypothesis*. A hypothesis is a very general statement made to tie together observational data prior to experimentation.

An *experiment* requires a set of procedures designed to provide data about a well-defined problem. The purpose of an experiment is to answer a question truthfully and accurately. A reliable experiment requires experimental conditions and control procedures. A *control* is a check on a scientific experiment and differs from the experimental procedure in one condition only. An experiment is designed to test a specific hypothesis.

A nontestable hypothesis is not useful in problem solving because it does not lend itself to experimental conditions. An example of a nontestable hypothesis follows: blow flies think that meat is an excellent site for egg laying. Because there is no way to test the thought processes of a blow fly, the hypothesis is invalid.

A classical work illustrating the effectiveness of a well-designed experiment is *Redi's Blow Fly Experiment*. That living things arise from lifeless matter was a popular 17th century belief. In 1650, the Italian physician Francesco Redi designed and performed an excellent experiment which disproved the spontaneous generation of blow flies from decaying meat:

In each of three jars, Redi placed the same kind and quantity of meat. The first jar was left uncovered. The second jar was covered with a porous cheese cloth. The third jar was covered with a parchment thick enough to prevent the odor of the meat from diffusing to the air outside of the jar (Fig. 1.1). Flies laid their eggs on the meat in the open jar and on the cheese cloth which covered the second jar. Eggs were not laid on the parchment covering the third jar. Redi watched as the eggs hatched into larvae (maggots) and observed the change of the larvae into adult flies. There were neither maggots nor flies on the meat in the third jar.

On the basis of this experiment, Redi concluded that maggots or flies do not come from decayed meat, but from the fertilized eggs laid by the female fly.

Fig. 1.1   Redi's Blow Fly Experiment. Can you explain the results?

When an hypothesis is supported by data obtained by experimentation, the hypothesis becomes a *theory*. A theory that withstands the test of time and further experimentation becomes a law or principle. A *concept* is a blanket idea covering several related principles.

THE TOOLS OF THE BIOLOGIST

Problem solving by the modern biologist requires the use of some very special tools and laboratory techniques. The scientific method is a major idea of biology upon which all research is built. All research begins with observation, a behavior that requires training, skill and concentration. To increase the accuracy of observation, the research biologist uses devices and techniques that will extend his or her vision and other senses. Because human sensing ability is limited and variable and not really reliable for seeing and recording fine and sometimes fleeting detail, modern scientists make use of devices and methods that enable them to perceive, measure and record data precisely.

Measurement is a basic requirement of all research. An investigator must answer questions concerning quantity, length, thickness, periods of time, mass, weight and the like within a hair's breadth of accuracy. Modern instruments of measurement permit astounding degrees of numerical precision. *Biometrics* is the science that combines mathematics and statistics needed to deal with the facts and figures of biology. Biologists handle enormous numbers which must be organized and simplified so that they become useful in the analysis of data. Computers aid the investigator's problem-solving tasks by accepting massive quantities of numerical data and making calculations at lightning speeds.

The laboratory or field scientist derives numerical data from using all kinds of measuring instruments designed for remarkable precision. Weighing in micro quantities requires the use of an *analytical balance* that is capable of weighing accurately to infinitessimal fractions of micrograms. The *manometer* measures the uptake of gases involved in cellular respiration and photosynthesis. *Spectrophotometers* are able to measure differences in densities of fluids by color comparisons not discernible to the human eye. Radioactivity is located and measured by the *Geiger counter*, a machine structured to feed pulses of electricity into an electronic counter. *Scintillation counters* measure light flashes imperceptible to the eye while ultraviolet rays are used to measure objects thinner than onion membrane.

Discovery in biology involves the ability to take apart the substances of cells so that the secrets of their biochemical processes can be exposed. The technique of *chromatography* permits the biologist to separate unbelievably small quantities of a substance into its component parts. For example: $\frac{1}{100,000}$ of a gram of protein can be separated into its constituent amino acids in one hour. Before the discovery of chromatography, a scientist needed 100 grams of protein and one year's time in order to make this separation. As the name implies, chromatography is "color writing," a technique that uses paper or a column of chalk as a stationary phase. The test sample, dissolved in a suitable solvent, is allowed to travel up the paper or through the chalk column. The constituents of the sample settle out at various levels as bands or dots of color (Fig. 1.2). *Electrophoresis* is a technique that uses electrical charges to separate the amino acids in proteins dotted on specially treated glass plates (Fig. 1.3).

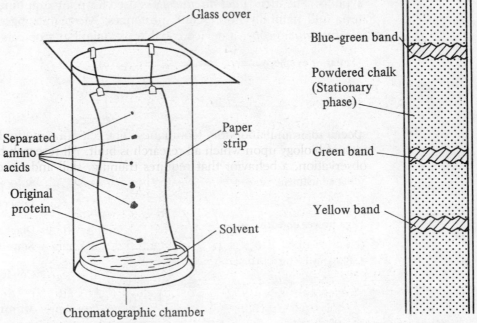

Fig. 1.2 Paper chromatography of a protein sample

Fig. 1.3 Column chromatography of green plant pigments

A biologist studying cell processes wishes to isolate the *organelles*, or parts, within the cell unit. When tissue and lysing (loosening) fluid are placed into a test-tube-like device and spun in a *centrifuge* at 60,000 revolutions per minute, the membranes enclosing the cells will split open. Based on weight differences, the organelles of each cell fall to the bottom of the tube at each successive spin of the centrifuge (Fig. 1.4).

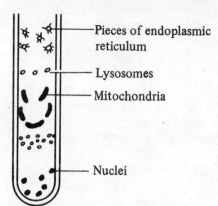

— Pieces of endoplasmic reticulum

— Lysosomes

— Mitochondria

— Nuclei

Fig. 1.4 Results of centrifugation—the heavier particles fall to the bottom of tube

A number of techniques and instruments have helped scientists see things better. *X-ray diffraction* is a procedure by which X rays sent through a crystal reveal the pattern of molecules and atoms contained in the crystal. This technique was used to determine the structure of the hereditary material DNA (deoxyribonucleic acid).

All sorts of microscopes have extended the investigator's field of vision. The best *light microscope* is capable of magnifying objects 2,000 times (Fig. 1.5). The *phase contrast microscope* makes transparent specimens visible, while the *darkfield* or the *ultramicroscope* gives vivid clarity to fragile and transparent organisms such as the spirochetes that cause syphilis. The *ultraviolet microscope* is used for photographing living bacteria and naturally fluorescent substances. Most amazing of all is the *electron microscope*, a device capable of magnifying objects more than

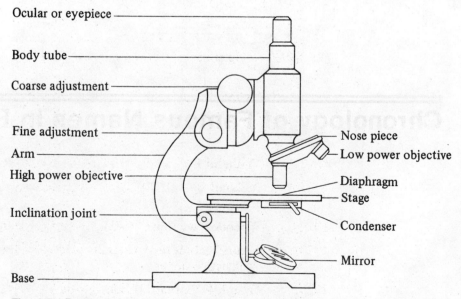

Ocular or eyepiece

Body tube

Coarse adjustment

Fine adjustment

Arm

High power objective

Inclination joint

Base

Nose piece

Low power objective

Diaphragm

Stage

Condenser

Mirror

Fig. 1.5 Parts of the compound (light) microscope

200,000 times. Using electrons instead of light and magnets in place of lenses, the electron microscope has revolutionized the study of the cell (Fig. 1.6). The *scanning electron microscope* has improved upon the resolution of fine detail made possible by electron microscopy. A fine probe directs and focuses electron beams over the material being studied, affording quick scanning and giving finer detail than is possible with the standard electron microscope.

Mention has been made herewith of only a small sampling of the kinds of tools and techniques available to the modern biologist. There are incubators for temperature control, refrigerators for keeping things cool, autoanalyzers for separating and classifying the elements in the blood, automatic mixers and stirrers and shakers, pH meters, autoclaves for sterilizing, timers and machines that measure time to the fraction of a second. The modern biologist uses anything and everything that helps to extend his or her senses and improve observation.

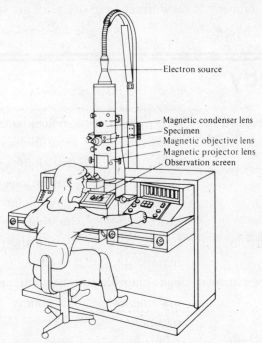

Electron source

Magnetic condenser lens
Specimen
Magnetic objective lens
Magnetic projector lens
Observation screen

Fig. 1.6   A biologist uses the electron microscope

Chronology of Famous Names in Biology

1895   **Wilhelm Konrad Roentgen** (German)—discovered X rays.

1903   **Mikhail Tsvett** (Russian)—discovered column chromatography.

1920   **Theodor Svedburg** (Swedish)—invented the centrifuge.

1923   **Georg von Hevesy** (English)—developed technique of using radioactive tracers.

1925   **Joseph E. Barnard** (English)—developed technique for photographing living bacteria using ultraviolet light.

1933	**Arne Tiselius** (Swedish)—developed electrophoresis.
1933	**Max Knell** and **Ernst Ruska** (German)—invented the single "lens" electron microscope.
1935	**Wendell Stanley** (American)—isolated the tobacco mosaic virus by centrifugation.
1938	**James Hillier** and **Albert Prebus** (English)—invented the compound electron microscope.
1940	**A.P.J. Martin** and **R.L.M. Synge** (English)—discovered the technique of paper chromatography.
1952	**Alfred D. Hershey** and **Martha Chase** (American)—developed the technique of using radioisotopes to study viruses.
1953	**Maurice H.F. Wilkins** (English)—developed the technique of using X-ray diffraction to decipher the structure of DNA.

Words for Study

analytical balance	embryology	molecular biology
anatomy	energetics	morphology
biology	evolution	paleobiology
biometrics	Geiger counter	paleontology
botany	genetics	physiology
centrifuge	histology	protoplasm
chromatography	homeostasis	scintillation counter
concept	hypothesis	spectrophotometer
controlled experiment	light microscope	spontaneous generation
cytology	manometer	taxonomy
ecology	naturalist	theory
electron microscope	nature study	zoology
electrophoresis		

Questions for Review

PART A. **Completion.** Write in the word that correctly completes each statement.

1. The science that specializes in the study of all life is ..1..
2. The ancient who founded the systemized study of natural history was ..2..
3. The first person to study the dissected human body was ..3..
4. The word biology was coined by ..4..

5. The branch of biology that specializes in the study of ancient life is ..5..

6. The science of classification is known is ..6..

7. Vital energy changes in the cell are studied in the science of ..7..

8. The study of plants and animals in their natural environment is known as the science of ..8..

9. Ichthyology is the study of ..9..

10. Steady state of the internal environment is described by the term ..10..

11. A tentative conclusion based on observation is the ..11..

12. Redi disproved the theory of ..12.. (2 words) of flies.

13. In the modern laboratory, the device used to measure the densities of colored liquids is the ..13..

14. A spinning machine used in the modern laboratory to separate cell structures is the ..14..

15. A device used to locate and measure radioactivity is the ..15..

PART B.   Multiple Choice. Circle the letter of the item that correctly completes each statement.

1. An example of a natural science is
   (a) geology   (c) physics
   (b) biology   (d) chemistry

2. The science of anatomy was founded by
   (a) Hooke   (c) Galen
   (b) Plato   (d) Tisellius

3. The "cells" in cork were named by
   (a) Hooke   (c) Galen
   (b) Plato   (d) Tisellius

4. The study of comparative anatomy was founded by
   (a) Malpighi   (c) Harvey
   (b) Lamarck   (d) Cuvier

5. The system of binomial nomenclature was developed by
   (a) Lamarck   (c) Leeuwenhoek
   (b) Linnaeus   (d) Lintel

6. The branch of biology that deals with function is
   (a) zoology   (c) physiology
   (b) taxonomy   (d) morphology

7. Included in the study of botany would be
   (a) dandelions   (c) guppies
   (b) sparrows   (d) skates

8. Inheritance and variation in living things is explained by the science of
   (a) ecology   (c) paleontology
   (b) genetics   (d) taxonomy

9. The study of insects is known as
   (a) ichthyology   (c) bryology
   (b) herpetology   (d) entomology

10. X-ray diffraction techniques were used to determine the structure of
    (a) ATP   (c) DNA
    (b) NAD   (d) PGA

11. The science of development is called
    (a) cytology   (c) systematics
    (b) embryology   (d) pteridology

12. A check on an experiment is a (an)
    (a) observation   (c) control
    (b) theory   (d) conclusion

13. A blanket idea covering several related principles is called a
    (a) theory   (c) concept
    (b) hypothesis   (d) conclusion

14. A microgram or its fraction can be weighed most accurately on a balance known as a (an)
    (a) analytical   (c) triple beam
    (b) spring   (d) double platform

15. Respiratory gases can be measured by using a device called a (an)
    (a) calorimeter   (c) barometer
    (b) manometer   (d) anemometer

PART C. Modified True-False. If a statement is true, write "true" for your answer. If a statement is incorrect, change the underlined word to one that will make the statement true.

1. A <u>botanist</u> is a scientist who specializes in the study of plants and animals.

2. The organized study of plants was founded by the ancient named <u>Hippocrates</u>.

3. The first person to see living cells was <u>Hooke</u>.

4. How blood circulates in the human body was first accurately demonstrated by <u>Harvey</u>.

5. Morphology is the branch of biology that studies <u>function</u>.

6. The study of tissues is known as <u>phylogeny</u>.

7. <u>Cytology</u> is the identification and ranking of plants.

8. The study of reptiles is known as <u>herpetology</u>.

9. <u>Ethology</u> is the science of birds.

10. The living substance of cells is <u>neoplasm</u>.

11. Problem solving in science is known as the scientific <u>module</u>.

12. A proven theory becomes a <u>conclusion</u>.

13. A combination of mathematics and statistics applied to facts of biology is known as <u>energetics</u>.

14. Amino acids in proteins can be separated by using electrical charges in a process known as <u>X-ray diffraction</u>.

15. The microscope that can magnify more than 200,000 times is the <u>electric</u> microscope.

Answers to Questions for Review

PART A

1. biology
2. Aristotle
3. Vesalius
4. Lamarck
5. paleobiology
6. taxonomy
7. energetics
8. ecology
9. fish
10. homeostasis
11. hypothesis
12. spontaneous generation
13. spectrophotometer
14. centrifuge
15. Geiger counter

PART B

1. b
2. c
3. a
4. d
5. b
6. c
7. a
8. b
9. d
10. c
11. b
12. c
13. c
14. a
15. b

PART C

1. naturalist or biologist
2. Theophrastus
3. Von Leeuwenhoek
4. true
5. form or structure
6. histology
7. Systematics
8. true
9. Ornithology
10. protoplasm
11. method
12. law or principle
13. biometrics
14. electrophoresis
15. electron

CHAPTER 2

CHARACTERISTICS OF LIFE

DEFINITION OF LIFE

As you already know, biology is the science that is concerned with life and living things. Given a collection of familiar objects, you would probably experience little difficulty in separating them into a group of living and a group of nonliving things. To determine if something is living, you would look for signs of life such as movement, response to touch, patterns of growth and ability to take in food. Let us suppose that included in the group of things you wish to classify are a mildew-covered towel, a green stain on tree bark, a large yellow turnip and slime mold on the forest floor. What signs of life would you look for in mildew, tree stain, the turnip and slime mold? Are these things living? What criteria are used to distinguish living organisms from nonliving objects?

Scientists are unable to agree on a standard definition of life because it is impossible to define life and its qualities in a sentence or two. For every definition that might be formulated, too many exceptions are found. For example, we can say that living things grow. But so can crystals be made to grow. They do so by adding on molecules of transparent quartz. We might also say that living things move. But what of the ball that rolls down a hill or a kite that flies in the air? Are these not moving? It is examples such as growing crystals and flying kites that make life a difficult concept to define. Life is best described in terms of the functions performed by living things.

Life Functions

Living things are highly organized systems. They are self-regulating and self-reproducing and capable of adapting to changes in the environment. To satisfy all of the conditions necessary for life, all living systems must be able to perform certain biochemical and biophysical activities which collectively are known as *life functions*.

Nutrition is the sum total of those activities through which a living organism obtains *nutrients* (food molecules) from the environment and prepares them for use as fuel and for growth. Included in nutrition are the processes of *ingestion, digestion* and *assimilation. Ingestion* is the taking in or procuring of food. *Digestion* refers to the chemical changes that take place in the body by which nutrient molecules are converted to forms usable by the cells. *Assimilation* involves the changing of certain nutrients into the protoplasm of cells.

Transport involves the absorption of materials by living things, including the movement and distribution of materials within the body of the organism. There are several transport methods, including diffusion, active transport and circulation. *Diffusion* is the flow of molecules from an area where these molecules are in great concentration to an area where there are fewer of them. *Active transport* is the movement of molecules powered by energy. *Circulation* is the movement of fluid and its dissolved materials throughout the body of an organism or within the cytoplasm of a single cell.

Respiration consists of breathing and cellular respiration. *Breathing* refers to the pumping of air into and out of the lungs of air-breathing animals or the movement of water over the gills of fish. During breathing, oxygen diffuses into the air sacs in the lungs and carbon dioxide moves out of the lungs through the nose and mouth. *Cellular respiration* is a combination of biochemical processes that release energy from glucose and store it in ATP (adenosine triphosphate) molecules.

Excretion removes waste products of cellular respiration from the body. The lungs, the skin and the kidneys are excretory organs in humans that remove carbon dioxide, water and urea from the blood and other body tissues. *Guttation* is the excretion of drops of water from plants during periods of high humidity.

Synthesis involves those biochemical processes in cells by which small molecules are built into larger ones. As a result of synthesis amino acids, the building blocks of proteins, are changed into enzymes, hormones and protoplasm.

Regulation encompasses all processes that control and coordinate the many activities of a living thing. Chemical activities inside of cells are controlled by enzymes, coenzymes, vitamins, minerals and hormones. The *nervous* and *endocrine systems* of higher animals integrate and coordinate body activities. Growth and development of plants is regulated by *auxins* and other growth-control substances.

Growth describes the increase of cell size and increase of cell numbers. The latter process occurs when cells divide in response to a sequence of events known as *mitosis.*

Reproduction is the process by which new individuals are produced by parent organisms. Basic to the understanding of reproduction is the concept that organisms produce the same kind of individuals as themselves. There are two major kinds of reproduction: asexual and sexual. *Asexual reproduction* involves only one parent. The parent may divide and become two new cells, thus obliterating the parent generation. Or the new individual may arise from a part of the parent cell; in such a case, the parent remains. In either case, *replication* (duplication) of the chromosomes is involved. *Sexual reproduction* requires the participation of two parents, each producing special reproductive cells known as *sex cells*, or *gametes.* The continuation and survival of the species is dependent upon reproduction. Once a species has lost its *reproductive potential*, the species no longer survives and it becomes extinct.

BASIC CONCEPTS IN BIOLOGY

Metabolism is an inclusive term concerning all of the biochemical activities carried on by cells, tissues, organs and systems necessary for the sustaining of life. Metabolic activities in which large molecules are built from smaller ones or in which nutrients are changed into protoplasm are called *anabolic activities*, or *anabolism*. Destructive metabolism in which large molecules are degraded for energy or changed into their smaller building blocks is called *catabolic activity*, or *catabolism*.

Homeostasis, as mentioned earlier, is a fundamental concept of biology. Homeostasis literally means "staying the same." Cells within the body of an organism must keep the chemical composition of the fluid contained in them and the chemistry of the water which surrounds them stable. To stay alive cells must regulate their internal and external fluids in temperature, acid-base composition, and the amount and content of mineral salts, ions and other substances that affect the steady state of their environment. This means that the life functions of an organism (or cell) are carried out in an integrated manner that helps the maintenance of a nearly unchanging internal environment.

Fig. 2.1 The long neck of the giraffe is an adaptation

Adaptation refers to a trait that aids the survival of an individual or a species in a given environment. An adaptation may be a structural characteristic such as the hump of a camel, a behavioral characteristic such as the mating call of a bull frog, or a physiological characteristic controlling some inner workings of tissue cells (Fig. 2.1). Adaptations permit the survival of species in environments that sometimes seem forbidding. For example, some bacteria are able to live in hot springs that have temperatures up to 80° C (175° F). They have adaptations that permit the carrying out of metabolic functions at extremely high temperatures.

How Living Things Are Named_____

THE CONCEPT OF SPECIES

The basic unit of classification is the *species*. A species is a group of similar organisms that can mate and produce fertile offspring. The red wolf, the African elephant, the red oak, the house fly, the hair cap moss—each belong to a separate and distinct species. For example, the red wolf belongs to the red wolf species in which the male mates with the female and produces fertile red wolf offspring. Upon maturity these red wolf offspring will reproduce just as their parents did. Species is a reproductive unit, not one defined by geography. Once brought together, a Mexican male Chihuahua can mate with a female Chihuahua born in France because they are species compatible.

Fig. 2.2 The mule, an infertile hybrid

In rare instances, members of closely related species—horses and donkeys, for example—can mate and produce offspring. But the products of interspecies matings are not fertile and therefore cannot reproduce. When a male donkey mates with a female horse (mare), a *mule* is produced. The mule is an infertile *hybrid* (Fig. 2.2). The mating of a male horse (stallion) and a female donkey results in a *hinny*, also an infertile hybrid.

BINOMIAL NOMENCLATURE

In modern biology living things are named according to the groups in which they are classified. The Swedish botanist and physician Carl von Linné (1707–1788) is credited with having devised the first orderly system of classification. Under the Latinized form of his name—*Carolus Linnaeus,* he is now known as "the father of modern taxonomy." Linnaeus based his system of classification on the anatomy and structure (morphology) of organisms. He recognized that living things fall into separate groups distinguishable by specific structural characteristics. He called each separate group of organisms a *species.* To each species he assigned a two-word Latin name or a *binomial.*

The first word in the species binomial designates the *genus.* Linnaeus assigned the name genus (plural *genera;* adjective *generic*) as the next higher unit of classification after the species. In his system the genus is a classification group consisting of one or more related species.

The species name is the *scientific name,* a double name in Latin known as *binomial nomenclature.* The system of double naming is so orderly and practical that it is used in every country in the world. Thus the scientific name for organisms is the same worldwide no matter the language of the country.

The scientific or species name—*Quercus rubra,* for example—provides us with some information about the organism. First of all, the genus name *Quercus* is the Latin name of oak. Species that are classified as oaks are grouped into the genus *Quercus.* The name *rubra* (red) indicates that it is a separate and distinct kind of oak apart from *Quercus alba* (white oak) or *Quercus suber* (cork oak). See Table 2.1.

TABLE 2.1. The Genus *Quercus*

Scientific Name	Common Name
Quercus alba	white oak
Quercus coccinea	scarlet oak
Quercus montana	chestnut oak
Quercus rubra	red oak
Quercus suber	cork oak
Quercus virginiana	live oak

Certain rules are followed in the use of the genus and species names. The genus name is capitalized; the second word in the species name is not. The scientific name (genus and species name) is always italicized in print and underlined when written in long hand. Whenever the species name is used more than once in the same paragraph, it can be abbreviated. The first letter of the genus name is used followed by the complete name of the species: *Q. rubra.* When used alone, the genus name is always capitalized and spelled out: *Quercus.*

TAXONOMIC GROUPS

The design of the classification system is a simple and practical one that easily lends itself to the addition of new names of organisms as they are discovered. The scheme makes use of a number of hierarchical groups each related to the other in terms of biological significance. Each group of organisms within the scheme is known as a *taxon* (plural, *taxa*). The classification groupings are as follows: *kingdom*, the largest and most inclusive group, followed by the *phylum, class, order, family, genus,* and *species*.

Linnaeus based his system of classification on the morphology (structure) of organisms in which those individuals that appeared most closely related were grouped into species and similar species into genera. The current practice is to group living things according to evolutionary relationships. *Evolution* is that branch of biology that studies changes in plants and animals that occur over long periods of time. Since the anatomical structures of plants and animals seem to reflect their evolutionary histories, the Linnaean system is compatible with the present day approach to classification.

TABLE 2.2. Classification of the Human

Category	Taxon	Characteristics
Kingdom	Animalia	Multicellular organisms requiring preformed organic material for food, having a motile stage at some time in its life history.
Phylum	Chordata	Animals having a notochord (embryonic skeletal rod), dorsal hollow nerve cord, gills in the pharynx at some time during the life cycle.
Subphylum	Vertebrata	Backbone (vertebral column) enclosing the spinal cord, skull bones enclosing the brain.
Class	Mammalia	Body having hair or fur at some time in the life; female nourishes young on milk; warm-blooded; one bone in lower jaw.
Order	Primates	Tree-living mammals and their descendants, usually with flattened fingers and nails, keen vision, poor sense of smell.
Family	Hominidae	Bipedal locomotion, flat face, eyes forward, binocular and color vision, hands and feet specialized for different functions.
Genus	* Homo	Long childhood, large brain, speech ability.
Species	** Homo sapiens	Body hair reduced, high forehead, prominent chin.

* Homo = Latin: man
** sapiens: wise

Five Kingdom System of Classification ___

The largest and most inclusive classification category is the kingdom. For many decades, living things were considered to be either plants or animals and thus were grouped into one of the two established kingdoms. Improved microscopic and other research techniques have revealed many species that cannot be classified either as plant or animal. For example: the one-celled organism *Euglena,* which has both plant and animal characteristics, was commonly referred to as a "biological puzzle" because taxonomists could not classify it to meet the satisfaction of all biologists. Other organisms such as bacteria, in which plant and animal characteristics could be demonstrated among the various species, also presented classifications problems.

In 1969, the ecologist Robert Whittaker proposed an updated system of classification in which living things are grouped into one of *five* kingdoms, based on the extent of their complexity and the methods by which their nutritional needs are met.

TABLE 2.3.   The Five Kingdom System

Kingdom	*Characteristics*
Monera	All monera are single-celled organisms. Unlike other cells, the monerans lack an organized nucleus, mitochondria, chloroplasts and other membrane-bound organelles. They have a circular chromosome. Examples are bacteria and blue-green algae.
Protista	Protists are one-celled organisms that have a membrane-bound nucleus. Within the nucleus are chromosomes that exhibit certain changes during the reproduction of the cell. Other cellular organelles are surrounded by membranes. Some protists take in food; other make it by means of *photosynthesis.* Some protists can move from one place to another (motile); others are nonmotile. Examples of protists are *Amoeba, Euglena, Paramecia, Punnularia* (diatom).
Fungi	Fungi are nonmotile, plant-like species that cannot make their own food. The fungi (fungus, singular) absorb their food from a living or nonliving organic source. Fungi differ from plants in the composition of the cell wall, in methods of reproduction and in the structure of body. Examples of fungi are mushrooms, water mold, bread mold.
Plantae	The plant kingdom includes the mosses, ferns, grasses, shrubs, flowering plants and trees. Most plants make their own food by photosynthesis and contain chloroplasts. All plant cells have a membrane-enclosed nucleus and cell walls that contain cellulose.
Animalia	All members of the animal kingdom are multicellular. The cells have a discrete nucleus that contains chromosomes. Most animals can move and depend on organic materials for food. Excluding the very simple species, most animals reproduce by means of egg and sperm cells.

Chronology of Famous Names in Biology

342 B.C.	**Aristotle** (384–322 B.C.)—first to try to group organisms by selecting a single outstanding feature.
1627–1705	**John Ray** (English)—introduced the word "species" to describe an organism. Catalogued about 19,000 plants of Europe.

1707–1778   **Carolus Linnaeus** (Swedish)—devised the first classification system. Wrote *Systems Natural* and *Species Plantarum*. Introduced the idea of the *type species*—typical specimen of the species.

1969   **Robert H. Whittaker** (American)—devised the five kingdom system of classification.

Words for Study

active transport	hybrid
adaptation	ingestion
anabolism	kingdom
animalia	life function
asexual reproduction	metabolism
assimilation	mitosis
binomial nomenclature	monera
breathing	mule
catabolism	nutrition
cellular respiration	order
class	phylum
circulation	plantae
diffusion	protista
digestion	regulation
excretion	replication
family	reproduction
fungi	respiration
gametes	sexual reproduction
genus	species
guttation	synthesis
growth	taxon
homeostasis	taxonomy
Homo sapiens	transport
hinny	

Questions for Review

PART A.   Completion. Write in the word that correctly completes each statement.

1. The taking in of food is known as ..1..
2. The type of transport requiring energy is ..2.. transport.

3. The pumping of air into and out of the lungs is known as ..3..

4. Small molecules are built into larger ones in biochemical processes known as ..4..

5. Examples of excretory organs are the kidneys, the skin and the ..5..

6. The building blocks of proteins are ..6..

7. Reproduction that involves one parent is known as ..7..

8. Chromosomes are duplicated in a process called ..8..

9. Destructive metabolism is known as ..9..

10. "To stay the same" is expressed by the biological term ..10..

11. The mating of a male donkey and a mare produce the animal known as a ..11..

12. Binomial nomenclature refers to ..12.. naming.

13. *Equus asinua* and *Equus caballus* indicate that these organisms belong to the same ..13..

14. The ..14.. name is always capitalized.

15. The science of classification is known as ..15..

PART B.   Multiple Choice. Circle the letter of the item that correctly completes each statement.

1. Chemical changes which change large food molecules into smaller soluble ones are known as
   (a) egestion   (c) mastication
   (b) ingestion   (d) digestion

2. The general term applied to the absorption and distribution of molecules within the body of an organism is
   (a) diffusion   (c) cyclosis
   (b) transport   (d) motion

3. Energy-releasing activities carried out by cells are known as
   (a) guttation   (c) digestion
   (b) respiration   (d) assimilation

4. The wastes of cellular respiration are
   (a) CO_2 and nitrogen   (c) carbon dioxide and water
   (b) urea and oxygen   (d) urea and nitrogen

5. Auxins are most correctly associated with
   (a) bean plants   (c) snails
   (b) giraffes   (d) beetles

6. Loss of water by plants during periods of high humidity is known as
   (a) excretion   (c) cremation
   (b) transportation   (d) guttation

7. All of the biochemical activities in the body are included in the term
   (a) anabolism   (c) metabolism
   (b) atavism   (d) catabolism

8. When species lose their reproductive potential, they
   (a) become extinct   (c) develop alternate systems
   (b) increase in numbers   (d) develop more energy

9. The long neck of a giraffe is considered to be a (an)
   - (a) throw-back
   - (b) affectation
   - (c) effector
   - (d) adaptation

10. The first word in the species binomial refers to the
   - (a) kingdom
   - (b) phylum
   - (c) genus
   - (d) class

11. The basic unit of classification is the
   - (a) taxon
   - (b) phylum
   - (c) kingdom
   - (d) species

12. A mule can properly be described as a
   - (a) high breed
   - (b) hybrid
   - (c) mongrel
   - (d) hinny

13. The name Linnaeus is best associated with
   - (a) taxonomy
   - (b) morphology
   - (c) paleobiology
   - (d) anatomy

14. The scientific name is the same as the
   - (a) common name
   - (b) species name
   - (c) genus name
   - (d) popular name

15. *Euglena* is best classified as a
   - (a) plant
   - (b) animal
   - (c) fungus
   - (d) protist

PART C. **Modified True-False.** If a statement is true, write "true" for your answer. If a statement is incorrect, change the underlined word to one that will make the statement true.

1. A definition of life is quite <u>easy</u> to formulate.

2. Living things are highly <u>dispersed</u> systems.

3. The process in which nutrients are changed into protoplasm is known as <u>digestion</u>.

4. The movement of materials from place to place in the body is known as <u>locomotion</u>.

5. Processes that involve control and coordination of the activities of living organisms are known collectively as <u>metabolism</u>.

6. Increase in cell size is known as <u>growth</u>.

7. The endocrine system is a part of the <u>plant</u> body.

8. Auxins help regulate growth in <u>plants</u>.

9. Two parents are required for the process of <u>asexual</u> reproduction.

10. Anabolism is a <u>breaking down</u> process.

11. A hinny <u>can</u> reproduce others like itself.

12. The <u>genus</u> name is a binomial.

13. The largest classification group is the <u>species</u>.

14. The branch of biology that studies change in form of living things is <u>classification</u>.

15. A classification group of organisms is known as a <u>taxon</u>.

Answers to Questions for Review

PART A

1. ingestion
2. active
3. breathing
4. anabolism
5. lungs
6. amino acids
7. asexual
8. replication
9. catabolism
10. homeostasis
11. mule
12. double (scientific)
13. genus
14. genus
15. taxonomy

PART B

1. d	6. d	11. d
2. b	7. c	12. b
3. b	8. a	13. a
4. c	9. d	14. b
5. a	10. c	15. d

PART C

1. difficult
2. organized
3. assimilation
4. transport or circulation
5. regulation
6. true
7. animal
8. true
9. sexual
10. building up
11. cannot
12. species
13. kingdom
14. evolution
15. true

CHAPTER 3

THE CELL—BASIC UNIT OF LIFE

The body of a living organism is built of units called *cells*. All living things are similar in that they are composed of one or more cells. The body of a *unicellular* organism is composed of one cell. Most plants and animals are *multicellular*, having a body made of numerous cells.

During the years 1838–1839, the *cell theory* was formulated by two eminent scientists of the day. Matthias Schleiden, a botanist, and Theodor Schwann, a zoologist, put together some of their fundamental ideas about the structure of plants and animals in what has developed into a basic concept of biological thought. The cell theory states (1) that cells are the basic units of life; (2) that all plants and animals are made of cells; and (3) that all cells arise from preexisting cells.

The Cell as a Basic Unit

UNIT OF STRUCTURE

Microscopic examination of plant and animal parts indicates that the bodies of living organisms are composed of cells. Cells provide structure and form to the body. They appear in a variety of shapes: round, concave, rectangular, elongate, tapered, spherical and other. Cell shape seems to be related to specialized function (Fig. 3.1).

Cells not only vary in shape, they also differ in size. Most plant and animal cells are quite small, ranging in size between 5 and 50 micrometers in diameter (Fig. 3.2). Cells are measured in units that are compatible with modern microscopes.

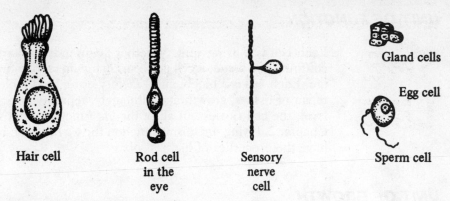

Fig. 3.1 Types of human cells

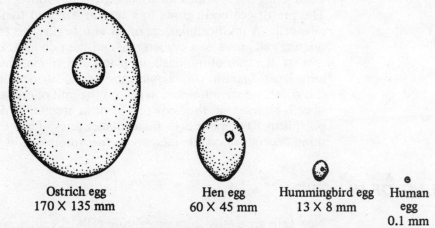

Fig. 3.2 Comparative sizes of egg cells

TABLE 3.1. Size Units Used In Microscopy

	Unit	Symbol	Equivalent	Measurement Use
Naked Eye Measurements	Meter	m	100 cm	Naked eye measurement; standard from which microscopic measurements are derived.
	Centimeter	cm	0.01 m; 0.4 in. 10 mm	Naked eye measurement of giant egg cells.
	Millimeter	mm	0.1 cm	Naked eye measurement of large cells.
Microscope Measurements	Micrometer **(Micron)	μm μ*	0.001 mm	Light microscope measurement of cells and larger organelles.
	Nanometer **(Millimicron)	nm mμ	0.001 μm	Electron microscopy measurement; cell fine structure; large macromolecules.
	**Angstrom unit	Å	0.1 nm	Electron microscopy measurement; molecules and atoms; X-ray methods.

* Pronounced *mew*.
** These units of measure are being phased out, but still appear in scientific literature.

UNIT OF FUNCTION

Each cell is a living unit. Whether living independently as a protist or confined in a tissue, a cell performs many metabolic functions to sustain life. Each cell is a biochemical factory using food molecules for energy, repair of tissues, growth and ultimately, reproduction. On the chemical level, the cell carries out all of the life functions that were discussed in Chapter 2. Living organisms function the way they do because their cells have the properties of life.

UNIT OF GROWTH

Each living thing begins life as a single cell. Protists remain unicellular. The protist cell body grows to a certain size and then divides into two new cells. A multicellular organism also begins life as a single cell. Its original cell grows to a certain size and then divides. Unlike the cells of protists, the cells of multicellular plants and animals hold together form- ing *tissues*. Later in this chapter, the role of tissues will be discussed. At this point, we are interested in cells as the unit of growth. As the number of cells increases in the body of a plant or animal, so does its size. Large organisms have more cells than smaller ones. Thus the size of a living thing depends upon the increase in the number of its cells.

UNIT OF HEREDITY

New cells arise only from preexisting cells. A cell grows to optimum size and then divides, producing two other cells *identical* to itself. Paramecia produce other paramecia; onion membrane cells produce new onion membrane cells identical in structure and function to themselves. From single cells, multicellular organisms produce other organisms like them- selves. This is so because cells carry hereditary information from one generation to the next. The information is coded in molecules of DNA (deoxyribonucleic acid) and is usually transmitted accurately to new cells. The reproductive machinery in the nucleus of the cell serves as a carrier of hereditary instructions.

Parts of a Cell

Essentially, all cells have similar parts designed to contribute to the work of the whole cell. The cell may be described as a membrane-enclosed unit consisting of *cytoplasm* and a *nucleus*. Both nucleus and cytoplasm are packed with finer structures that can be resolved or distinguished clearly only through the electron microscope.

The parts of a cell are known as *organelles*, meaning "little organs." This term is appropriate because parts of the cell have special functions, somewhat like miniature body organs. Fig. 3.3 shows a typical animal cell as demonstrated through the light microscope. However, there are many organelles in the cell that cannot be seen by using the light mi-

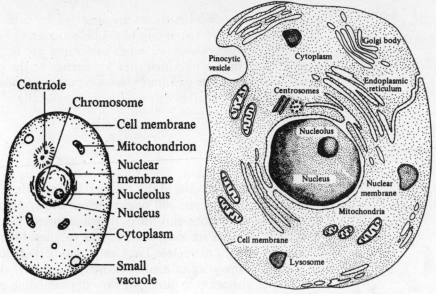

Fig. 3.3 Typical animal cell

Fig. 3.4 Electron micrograph of a cell

croscope. These structures can be resolved by the electron microscope and are known collectively as the *fine structure of the cell*. Fig. 3.4 presents an *electron micrograph* of a cell. You will find it helpful to refer to this diagram as you read about the structure and functions of parts of the cell.

CELL MEMBRANE

The outer boundary of the cell is called the *cell* or *plasma membrane*. Although it is no thicker than 10 nanometers, the cell membrane has an intricate molecular structure. It is composed of a bilayer of phospholipid molecules with proteins of various sizes embedded in the layers. *Pores* in the cell membrane are lined with small protein molecules.

The cell membrane controls the passage of materials into and out of the cell. It is often referred to as a living gatekeeper. The cell membrane is *semi-permeable* and highly selective: not every ion or molecule can cross its boundary. The movement of materials across the cell boundary and into or out of the cell is given the general term of *transport*. It is controlled by the globular proteins, the phospholipids and the pores of the membrane and by the electrochemical nature of *protoplasm*, the living substance of the cell.

Overall, there are two major types of transport: *passive transport* and *active transport*. Passive transport does not require the cell's energy to move molecules. By the process of *diffusion* molecules move from an area of greater concentration to an area of lesser concentration. *Osmosis* is the movement of water across a semi-permeable membrane. *Plasmolysis* is the shrinking of cytoplasm due to the movement of water out of the cell. Both osmosis and plasmolysis are forms of passive transport.

Active transport requires an expenditure of energy by the cell. *Pinocytosis*, or "cell drinking," is a form of active transport by which fluid molecules are engulfed by cells through the formation of vesicles in the cell membrane. Solid particles are ingested by cells through a process known as *phagocytosis*. White blood cells ingesting bacteria serve as an example of phagocytosis. At times molecules are forced out of cells by *exocytosis*, a means by which they are carried to the cell surface by vacuoles or vesicles. The *sodium-potassium pump* is a means by which excess sodium ions are forcibly extruded from nerve cells while potassium ions are pulled into the cell.

CYTOPLASM

All of the living material of the cell that lies outside of the nucleus is the *cytoplasm*. It is packed full of organelles and is highly structured by a network of fine tubes and fibers that are spread throughout the entire cell. *Microtubules* are long, thin, hollow tubules (little tubes), measuring 25 nanometers in diameter. They are composed of globular proteins and act as the framework of a submicroscopic internal skeletal system. Microtubules also seem to direct the flow of circulating materials within the cytoplasm. *Microfilaments* are long protein threads that measure about 6 nanometers in diameter. They function in cell movement.

ENDOPLASMIC RETICULUM

Spreading throughout the cytoplasm, extending from the cell membrane to the membranes of the nucleus is a network of membranes that form channels, tubes and flattened sacs; this network is named the *endoplasmic reticulum*. One function of the endoplasmic reticulum is the movement of materials throughout the cytoplasm and to the plasma membrane. The endoplasmic reticulum has other important functions related to the synthesis of materials and their packaging and distribution to sites needed. Some membranes of the endoplasmic reticulum are dotted with thousands of organelles known as ribosomes; others are smooth.

RIBOSOMES

Ribosomes are small, circular organelles measuring 25 nanometers in diameter. They are by far the most numerous organelles in a cell. For example, in one bacterial cell ribosomes may number upward of 15,000. Some ribosomes are attached to the membranes of the endoplasmic reticulum. These are engaged in the synthesis of proteins that will be exported from the cell to be used by various organs of the body. Digestive enzymes and hormones are examples of the kinds of protein molecules that are secreted by cells to be used elsewhere in the body. Ribosomes that lie free in the cytoplasm synthesize proteins that are to be used in the cell in activities such as cellular respiration. In general, ribosomes are the sites of protein synthesis.

GOLGI BODY

The Golgi body, sometimes referred to as the Golgi apparatus, consists of a series of membranes that are loosely applied to one another forming *vesicles* (fluid-filled pouches) that are surrounded by microtubules. The

Golgi body receives vesicles and their fluids from membranes of the endoplasmic reticulum. The vesicles are rewrapped in membranes by the Golgi body and transported to the cell membrane where they leave the cell. In general, the function of the Golgi body is to temporarily store, package and transport materials synthesized by other parts of the cell. Plant cells have several hundred Golgi bodies; animal cells, usually 10 to 20.

LYSOSOMES

The lysosomes are vacuoles bounded by double membranes which keep them isolated from other cellular organelles. The lysosomes contain hydrolytic enzymes that are capable of destroying the cell. By a well-controlled mechanism, lysosomes become attached to food vacuoles and release some of the digestive enzymes into the food vacuoles. Bacteria engulfed by white blood cells are digested in this way.

MITOCHONDRION

Aside from the nucleus, the mitochondrion is the largest organelle in the cell. Because there are so many of them, the plural mitochondria is used more frequently than the singular mitochondrion. In diameter, mitochondria measure between 0.5 and 1 micrometer; in length they measure up to 7 micrometers. Mitochondria, known as the "power houses of the cell," exhibit a variety of shapes: round, ovoid, elongated, and the like. Each mitochondrion is surrounded by a double membrane and has a membrane of many folds fitted into its internal structure. The internal membranes, *cristae*, actually increase the surface area, permitting a great amount of biochemical activity to take place. In addition to increasing surface area, the cristae form compartments providing additional work and storage areas for the complex task of *cellular respiration*. Depending on their energy needs, cells may have more than 2,500 mitochondria (Fig. 3.5).

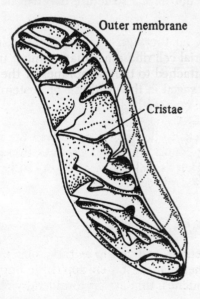
Outer membrane

Cristae

Fig. 3.5 Mitochondrion

NUCLEUS

The largest and most prominent organelle in the cell is the nucleus. A few primitive species such as bacteria and blue-green algae do not have an organized nucleus. These types of cells are known as *prokaryotes*. Most cells do, however, have a double membrane-bound nucleus and are classified as *eukaryotes*.

The double membrane of the nucleus fuses at certain places forming openings called *pores*. The pores measure about 65 nanometers in diameter. Inside the nuclear membrane a clear semi-solid material seems to fill up the nucleus. Embedded in this material are one or two small spherical bodies called *nucleoli* (singular, nucleolus). The nucleolus is the site of the synthesis and storage of the nucleic acid RNA (ribonucleic acid). In the nucleus of the nondividing cell is a tangle of very fine threads which absorb stain quite readily. In the granular stage these threads are known as *chromatin*. The chromatin threads come together, shorten and thicken forming *chromosomes* that can be seen quite prominently in the dividing cell.

Comparison of Plant and Animal Cells

A typical plant cell is shown in Fig. 3.6. Notice that plant cells have as their outer boundary a *cell wall* which surrounds the plasma membrane. The cell wall is a nonliving structure composed of *cellulose*, a complex starch molecule. Cellulose molecules are bound together in an intricate pattern and held in place by gluey carbohydrates known as *pectins*. The carbohydrate molecules that compose the cell wall are synthesized by the cytoplasm of the cell and secreted through the cell membrane to form a rigid boundary around the cell.

The cell wall serves to support the cell, to protect it from drying out and to inhibit bacterial invasion. If the cytoplasm inside the cell loses water and shrinks, the cell wall still retains its shape and remains fairly rigid. Animal cells do not have a structure that can be compared to the cellulose cell wall.

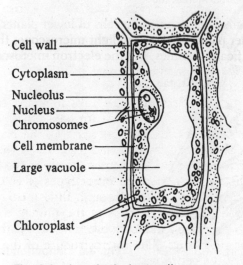

Cell wall
Cytoplasm
Nucleolus
Nucleus
Chromosomes
Cell membrane
Large vacuole

Chloroplast

Fig. 3.6 A typical plant cell

Fig. 3.6 shows a plant cell with a very large *vacuole*. A vacuole is a space in the cytoplasm filled with water and dissolved substances such as salts, sugars, minerals and other materials. The *cell sap* in the vacuoles of some cells contain pigments that color the flowers, fruits, leaves and stems of plants. Like other organelles, the vacuole is surrounded by a membrane which separates it from the cytoplasm of the cell.

Storage areas in animal cells are quite small, occupying very little space in the cytoplasm. These storage areas in animal cells are known as *vesicles*. They are neither fixed in number nor as lasting as organelles, but rather appear and disappear as they move materials from the endoplasmic reticulum to the Golgi body to the cell membrane.

Chloroplasts belong to a group of structures which has the general name *plastid*. Plastids are membrane-bound organelles found only in plant cells. Usually plastids are spherical bodies that float freely in the cytoplasm, holding pigment molecules or starch. Chloroplasts contain the green pigment *chlorophyll*, a substance that gives plants the green color. Chlorophyll is a special molecule that has the ability to trap light and to convert it to a form of energy that plants can use in carrying out the chemical steps of the food-making process known as *photosynthesis*.

Each chloroplast is surrounded by a double membrane. Inside the chloroplast are numerous flattened membranous sacs called *thylakoids* (formerly called *grana*). The thylakoids are the structures that contain the chlorophyll and it is within these sacs that photosynthesis takes place. *Stroma* is the name given to the dense ground substance that cushions the thylakoids. Animal cells do not have chloroplasts and therefore cannot make their own food. Fig. 3.7 shows the fine structure of a chloroplast.

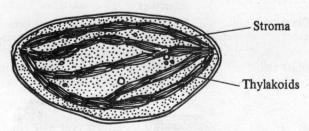

— Stroma

— Thylakoids

Fig. 3.7 A chloroplast

Centrioles are paired structures that lie just outside of the nucleus of nearly all animal cells and some cells of lower plants. They are absent in cells of higher plants. Under the light microscope, the centrioles look like two insignificant granules, but the electron microscope demonstrates that they have a very intricate structure (Fig. 3.8).

Flagella and *cilia* are fine threads of cytoplasm that extend from the surfaces of some cells. Both of these structures are involved in the locomotion of some protist species. Cilia are relatively short extensions but appear in great numbers, usually surrounding the body of the protist. Flagella are much longer than cilia and appear in fewer numbers.

Microtubules

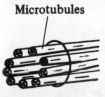

Fig. 3.8 The fine structure of a centriole

In addition to serving the locomotive needs of one-celled organisms, flagella and cilia help functions of other types of cells. Sperm cells of animals and plants are propelled through fluid media by the whip-like actions of their flagella. Tissue cells of the human windpipe are lined with cilia which wave back and forth catching dust particles and pushing them away from the lungs. The microstrucure of the flagella and cilia resembles that of the centrioles.

Organization of Cells and Tissues_____

Cells in the body of the multicellular organism are arranged in structural and functional groups called *tissues*. A tissue is a group of similar cells that work together to perform a particular function. Tissues that are grouped together and work for a common cause form *organs*. Groups of organs that contribute to a particular set of functions are called *systems*. The ability of cells to carry out special functions in addition to the usual work of cells exemplifies *specialization*. When different jobs are accomplished by the various tissues in an organ, we call this *division of labor*.

Tables 3.2 and 3.3 provide a summary of major animal and plant tissues.

TABLE 3.2.   Animal Tissues

Tissue Type	Function of Tissue
Epithelial	Made of closely packed cells specialized for covering and lining organs and protecting underlying tissues from drying out, mechanical injury and bacterial invasion (Fig. 3.9).
squamous cells	Flattened epithelial cells; specialized for lining body cavities such as the mouth, esophagus, eardrum and vagina.
cuboidal cells	Cube-like cells; specialized for gland tissue and the lining of the kidney tubules.
columnar cells	Tall and column-like cells; specialized for lining the alimentary canal in such organs as the stomach and intestines.
ciliated columnar cells	Ciliated columnar epithelial cells line the tubes that serve as air passageways in the respiratory system; the cilia push dust particles away from the lungs.
goblet cells	Specialized for secreting and storing fluid products of cells such as milk, hormones, enzymes and oils.

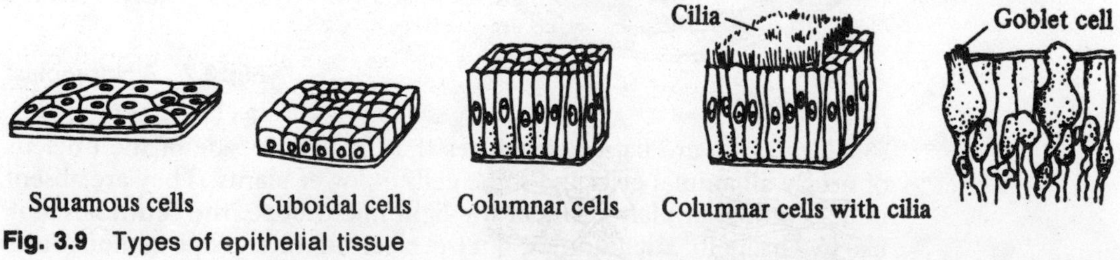

Squamous cells   Cuboidal cells   Columnar cells   Columnar cells with cilia

Fig. 3.9   Types of epithelial tissue

Nervous	Made of cells called *neurons* that are specialized for carrying impulses; the nervous system coordinates all of the body's activities (Fig. 3.10).
sensory neuron	Carries information from sense organs to other neurons in the brain and spinal cord.
motor neurons	Carries impulses to muscles and glands.
interneurons	Transmits impulses from sensory neurons to motor neurons.

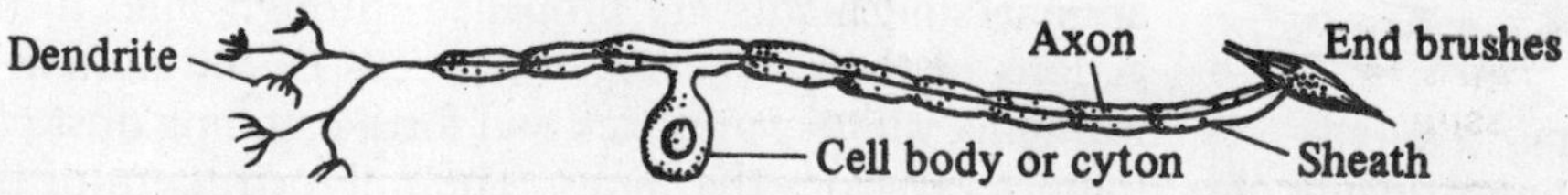

Fig. 3.10   Nerve cell

Tissue Type	*Function of Tissue*
Muscle	Specialized to respond to stimuli transmitted by motor neurons; characterized by electrical excitability and the ability to contract (Fig. 3.11).
smooth	Involuntary muscle; composes organs not under the control of the individual—e.g. small intestine.
cardiac	Specialized for the heart.
striated	Voluntary muscle; composes organs that are controlled by the will of the individual—e.g. arm and leg muscles.

Fig. 3.11 Types of muscle tissue

Smooth or involuntary muscle   Cardiac or heart muscle   Striated or voluntary muscle

Blood	Composed of three types of cells suspended in plasma (Fig. 3.12).
erythrocytes	Red blood cells specialized for carrying oxygen.
leucocytes	White blood cells specialized for counteracting invasions by disease organisms.
platelets	Cell fragments that function in blood clotting.

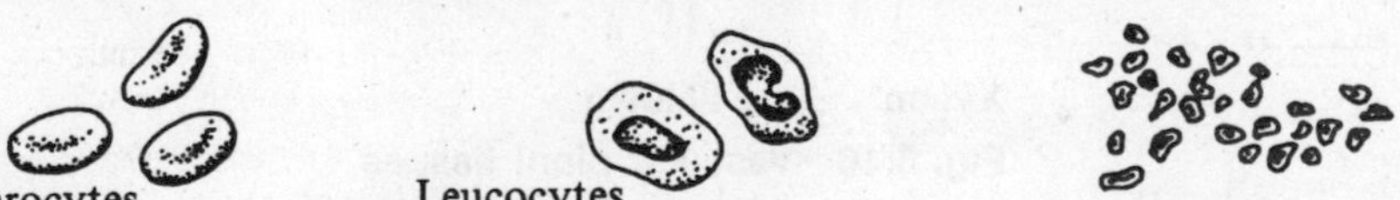

Fig. 3.12 Types of blood cells

Erythrocytes Red blood cells   Leucocytes White blood cells   Platelets

Connective	Characterized by large deposits of nonliving material that surrounds living cells; the nonliving *matrix* supports and binds tissues to the body skeleton (Fig. 3.13).
cartilage	Firm and elastic matrix found at tip of nose, end of long bones, ear lobes and wherever strength and flexibility are needed.
fibrous connective tissue	Parallel protein fibers give strength to the matrix; *ligaments*—elastic fibers—connect bone to bone. *Tendons*—nonelastic fibers—connect muscle to bone.

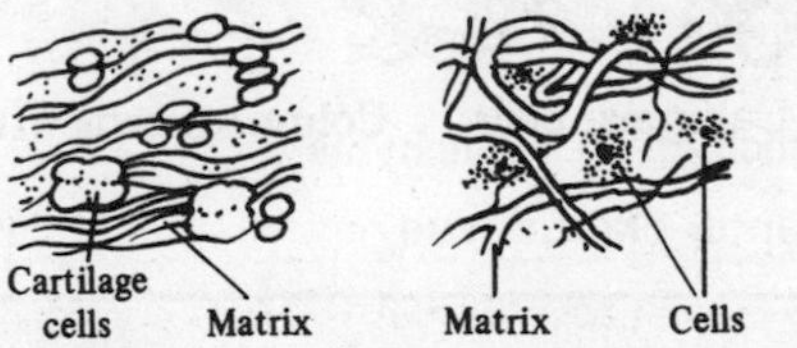

Fig. 3.13 Types of connective tissue

Cartilage cells   Matrix   Matrix   Cells
Cartilage   Fibrous connective tissue

Bone	Calcium and phosphorus compounds make up the matrix; living cells arranged in rings around a canal through which nerve fibers and blood vessels extend. The skeleton of vertebrates is made of bone (Fig. 3.14).

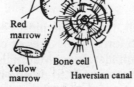

Red marrow
Yellow marrow   Bone cell
Haversian canal

Fig. 3.14 Bone tissue

TABLE 3.3.   Plant Tissues

Tissue Type	Function of Tissue
Epidermis	One cell-layer thick; covering surfaces of leaves, stems and roots, protecting inner tissues (Fig. 3.15).

Leaf epidermis   Cork cells

Fig. 3.15   Epidermal plant tissue

Vascular	Conducting tissues; transport materials throughout plant (Fig. 3.16).
xylem	Composed of elongated cells known as *tracheids* and *vessels*, or conducting tubes; conducts water and its dissolved minerals upward from roots of plant.
phloem	Transports food materials to all parts of the plant; composed of *sieve tubes* and *companion cells*.

Xylem   Phloem

Fig. 3.16   Vascular plant tissues

Fundamental	Basic tissues that make up most of the plant body (Fig. 3.17).
parenchyma	Thin-walled cells containing chloroplasts and other plastids; tissues where photosynthesis takes place.
sclerenchyma	Tough supporting tissues composed of sturdy cell walls and dead cells; gives mechanical support to plant stems and forms tough coverings of seeds.
collenchyma	Living cells in stems and leaves that provide support.

Parenchyma   Sclerenchyma   Collenchyma

Fig. 3.17   Fundamental plant tissue

The Basic Chemistry of Cells

Most of the activities of the cell involve chemical changes. Small molecules may be joined together to form larger ones or complex substances may be broken down into their smaller units. Materials are changed from one form to another, used up or synthesized during biochemical events that take place in cells.

CHEMICAL COMPOUNDS IN LIVING CELLS

The cell is likened to a "chemical factory" that uses some of the elements present in the nonliving environment. Of the elements in the living material of the cell, carbon, hydrogen, oxygen, and nitrogen are present in the greatest amounts. Sulfur, phosphorus, magnesium, iodine, iron, calcium, sodium, chlorine and potassium are found in smaller quantities. These elements are present in inorganic and organic compounds that are utilized by the cell.

Inorganic compounds are compounds that do not have the elements carbon and hydrogen in chemical combination. The inorganic compounds in greatest percentages in cells are water, mineral salts, inorganic acids and bases. For example: hydrochloric acid (HCl), an inorganic acid, is secreted by the gastric glands in the stomach, acidifying stomach contents.

Organic compounds are compounds that contain the elements carbon and hydrogen in chemical combination. Organic compounds are produced by living plants and animals and can be synthesized in the laboratory, as well. The special bonding property of carbon permits it to form compounds that are structured as long chains of atoms or as rings of atoms. The types of organic compounds that are contained in the living material of the cell and used in its chemical activities are carbohydrates, lipids, proteins, and nucleic acids.

Carbohydrates

Carbohydrates are composed of the elements carbon, hydrogen, and oxygen. Hydrogen and oxygen atoms are usually present in carbohydrates in the ratio of 2:1. Glucose ($C_6H_{12}O_6$) represents the basic unit of carbohydrate structure, a *monosaccharide* molecule. A monosaccharide is a simple sugar from which larger carbohydrate molecules are built. When two monosaccharide molecules are joined together chemically, a double sugar or *disaccharide* is formed. As the two molecules join, a molecule of water is produced during the process in addition to the double sugar. A synthesis of this type is known as *dehydration synthesis*.

$$C_6H_{12}O_6 + C_6H_{12}O_6 \rightarrow C_{12}H_{22}O_{11} + H_2O$$

glucose + glucose → maltose + water

Within living cells, many monosaccharides chemically combine by dehydration synthesis and form a *polysaccharide*, such as cellulose.

Carbohydrates are used by cells primarily as sources of energy. In plant cells, polysaccharides are utilized in cell structures such as cell walls. In some animal cells, glycoproteins (carbohydrate and protein compounds) form recognition sites on the membranes of certain cells.

Lipids

The lipids are a group of organic compounds that include the fats and fat-like substances. A lipid molecule contains the elements carbon, hydrogen and oxygen like that of a carbohydrate. Unlike the carbohydrates, however, in lipid molecules the ratio of hydrogen to oxygen is much greater than 2:1. A lipid molecule is made up of two basic units: an *alcohol* (usually glycerol) and a class of compounds called *fatty acids*.

Lipids are sources for biologically usable energy. Most lipid molecules provide twice as much energy per gram as do carbohydrate molecules. However, the energy in lipid molecules is not so easy to extract as that of glucose. Besides being useful for energy, lipids are essential to the structure of cells. The plasma membrane is composed of lipid molecules as is the myelin sheath of some nerve fibers.

Proteins

Proteins are organic compounds composed of the elements carbon, hydrogen, oxygen and nitrogen. Some proteins also contain sulfur. All proteins are built from small molecular units known as *amino acids*. The amino acid molecules link together in a particular way through *peptide* bonds. A *dipeptide* consists of two amino acids. A *polypeptide* contains many amino acid molecules. A *protein* is composed of one or more polypeptide chains.

About 20 amino acids are essential to living systems. From these a large number of different kinds of proteins are formed. The great variety of protein molecules is possible because of the many ways in which amino acid molecules can be arranged. Changing the sequence of just one amino acid in a chain will change the protein molecule.

Much of the work of the cell is concerned with the synthesizing of protein molecules. Some of these proteins such as hormones, enzymes and hemoglobin are used in complex biochemical activities. Other proteins contribute to the structure of cells such as those that make up the plasma membrane and other cellular membranes.

Nucleic Acids

Nucleic acids are a group of organic compounds that are essential to life. These are the compounds that pass hereditary information from one generation to another, making possible a remarkable continuity of life within the various species of living things. Deoxyribonucleic acid (DNA) molecules are the particular type of nucleic acid out of which genes are made. Genes are the bearers of hereditary traits from parent to offspring.

Another important characteristic of nucleic acids is their ability to carry information from genes in the cell nucleus to certain structures in the cytoplasm which direct major biochemical processes. For example: the building of proteins is controlled by the group of nucleic acids known in general as ribonucleic acid (RNA). The structures and functions of the three types of RNA molecules and the structure and function of DNA are presented in detailed discussion in Chapter 14.

THE ROLE OF ENZYMES IN LIVING CELLS

Cells are always engaged in chemical activity. The major difference between living things and nonliving matter is that living systems carry out vital chemical activities on a continuous and controlled basis. The control of chemical processes in cells requires the work of *enzymes*.

Enzymes are *organic catalysts*. A catalyst is a molecule that controls the rate of a chemical reaction but is itself *not* used up in the process. Enzymes control the rate of chemical reactions that take place in cells, tissues and organs. Each chemical reaction that occurs in a living system requires the assistance of a specific enzyme.

The Structure and Nature of Enzymes

Enzymes are large complex proteins made up of one or more polypeptide chains. Enzymes may be entirely protein or they may have nonprotein parts known as *coenzymes*. Frequently, vitamins function as coenzymes. For example: riboflavin, popularly called vitamin B_2, is important in a coenzyme function during the cellular process of respiration. Scientists affix the *ase* ending to names of enzymes.

Enzymes have several characteristics that are important to the chemistry of living cells. These characteristics are due to their protein nature. First, enzymes are made inactive by heat. In the human body enzymes work best at normal body temperature of 98.6°F, or 37°C. An increase in temperature of only a few degrees can render enzymes inactive. Second, the action of enzymes can be blocked by certain compounds. We call these enzyme blocks *poisons*. Hydrogen cyanide blocks one of the enzymes that is involved in cellular respiration. Third, enzymes are specific in their activity. Usually they catalyze one particular reaction. For example: the enzyme maltase will only catalyze the digestion of the sugar maltose into two glucose molecules.

How Enzymes Work

Biochemical evidence provides insight into how enzymes work. The surface configuration of an enzyme is known as its *active site*. The *substrate* is the substance upon which the enzyme works. An enzyme affects the rate of reaction of the substrate molecule that fits the enzyme's activity site. In order for this to happen, a close physical association must take place between enzyme and substrate. This association is called the *enzyme-substrate complex* (Fig. 3.18). While the enzyme-substrate complex is formed, the internal energy state of the substrate molecule is changed, bringing about a chemical reaction. As soon as the reaction is completed, the enzyme and the products separate. Each enzyme works at its own maximum rate. The rate at which an enzyme can catalyze all of the substrate molecules that it can in a given time is called the *turnover number*. Different enzymes have different turnover numbers. However, the general rate of enzyme action is influenced by certain environmental conditions such as temperature and relative amounts of enzyme and substrate.

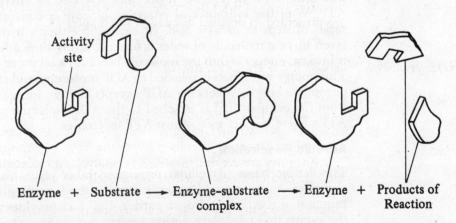

Activity site

Enzyme + Substrate ⟶ Enzyme–substrate complex ⟶ Enzyme + Products of Reaction

Fig. 3.18 Enzyme-substrate complex

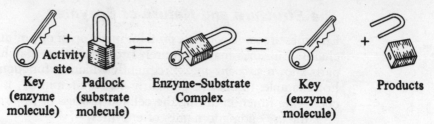

| Key (enzyme molecule) | Padlock (substrate molecule) | Enzyme–Substrate Complex | Key (enzyme molecule) | Products |

Fig. 3.19   Lock and key analogy of an enzyme and its substrate

A simple analogy is used to explain the *specificity* of enzymes. Specificity refers to the characteristic of enzymes which permits a particular enzyme to form a complex with a specific substrate molecule only. The "lock and key" analogy explains enzyme specificity: the substrate is viewed as a padlock and the enzyme as the key able to unlock it. When unlocked (in the analogy) or acted upon by the key, the padlock comes completely apart. The key remains unchanged and ready to work again on another padlock of the same type (Fig. 3.19).

CELLULAR RESPIRATION

Living cells require a constant supply of energy to fuel the chemical activities that sustain life. Glucose is the major supplier of the cell's energy. The cell is able to extract energy from glucose in small packets. The energy released is then stored in two molecules: *adenosine diphosphate* (ADP) and adenosine triphosphate (ATP).

Types of Cellular Respiration

Energy is produced by the cell during two major pathways: *anaerobic respiration*, also known as *glycolysis*, and aerobic respiration.

Anaerobic Respiration

As is implied by the name, oxygen is not used during anaerobic respiration. Fig. 3.20 provides a summary of the events that take place during this phase of cellular respiration.

Anaerobic respiration takes place in the cytoplasm of cells. It begins with a molecule of glucose. It becomes activated by energy supplied by ATP. With the assistance of enzymes, glucose is converted through a series of steps to *pyruvic acid*. At each step either a hydrogen atom is given up or a molecule of water is formed. Every time a hydrogen bond is broken, energy within the molecule becomes a bit more concentrated. This energy is ultimately released to ADP molecules and stored between phosphate bonds. Whenever ADP accepts energy, inorganic phosphate from the cellular fluid is attached to the ADP molecule upgrading it to ATP. There is a net gain of two ATP molecules.

Aerobic Respiration

The *aerobic phase* of cellular respiration takes place inside the mitochondria. As the name implies, oxygen is used, serving the important function of the final hydrogen carrier. Fig. 3.21 provides a summary of the events that take place during aerobic respiration, also known as the Krebs cycle.

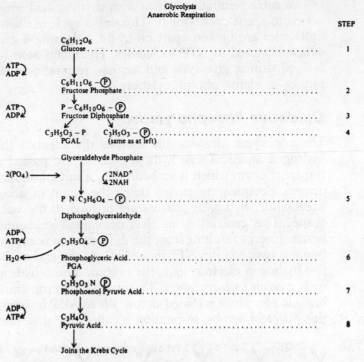

Fig. 3.20 Anaerobic respiration, or glycolysis

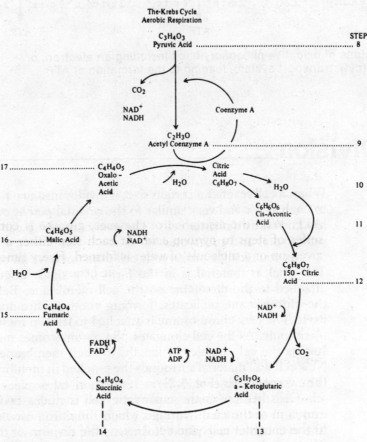

Fig. 3.21 Krebs citric acid cycle

Aerobic respiration begins with pyruvic acid which is quickly converted to acetyl coenzyme A. Through a cycle of chemical changes, fuel molecules are broken apart bit-by-bit, releasing a great deal of energy which is stored in ATP molecules. Hydrogen atoms from compounds formed during glycolysis and aerobic respiration then enter the next phase—oxidative phosphorylation.

Oxidative Phosphorylation

The coenzyme nicotinamide adenine dinucleotide (NAD) receives the hydrogen atoms. These hydrogen atoms are passed along an electron transport chain which is embedded in a mitochondrial membrane. Hydrogen electrons are passed down the chain of acceptors (Fig. 3.22). Meanwhile, hydrogen protons are pushed to the outside of the membrane. This establishes an electrochemical gradient across the membrane. Energy resulting from the difference in potential across the membrane is used to form ATP molecules. At the end of the transport chain, the hydrogen electron joins the proton. These hydrogen ions combine with oxygen to form water. During the electron transport process, inorganic phosphate is joined chemically to ADP forming ATP. Therefore, this phase of aerobic respiration is called oxidative phosphorylation.

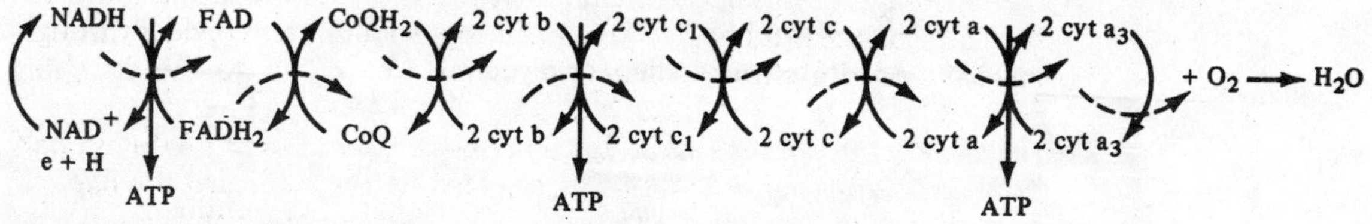

Fig. 3.22   Steps of oxidative phosphorylation including an electron, or cytochrome (cyt), transport system, leading to the formation of ATP

Cell Division

When a cell reaches a certain size, it divides into two new cells, identical to each other and very similar to the original *parent* cell. The new cells are known as *daughter* cells. The events marking cell division differ in prokaryotes and eukaryotes.

As you recall, a prokaryotic cell does not have an organized nucleus. The nuclear material is in the form of a single circular chromosome attached to the *mesosome* on the cell membrane. Before cell division, the chromosome replicates, forming another chromosome exactly like itself. The new chromosome is attached to its own mesosome on the cell membrane. As the cell elongates, the chromosomes move further apart. Just as the cell doubles in length, the cell membrane pinches inward. New cell wall material surrounds the pinched in membrane and separates the two new cells (Fig. 3.23).

Cell division in the eukaryotic cell includes two separate events: events in the nucleus through which the chromosomes are distributed to the daughter cells and *cytokinesis*, the division of the cytoplasm and its organelles.

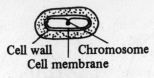

Fig. 3.23 Division of a bacterium

Each species has a characteristic number of chromosomes in the nuclei of its cells. It is referred to as the *species number*, or simply as the *chromosome number*. As indicated in Table 3.4, the number of chromosomes in the cells of a given species is not an indicator of size or complexity. The somatic, or body, cells of an organism contain the full complement of chromosomes, referred to as the *diploid number* and abbreviated by the symbol 2N. The nucleus of somatic cells divides through *mitosis* and the chromosomes are distributed equally to the daughter cells.

The organism's sex cells, or gametes—eggs and sperm, contain half the species number of chromosomes. This number is called the *haploid number* and abbreviated N. *Meiosis* is the kind of nuclear division that leads to the formation of sperm and egg cells.

TABLE 3.4. Chromosome Numbers of Some Common Species

Organism	Haploid No.	Diploid No.
mosquito	3	6
fruit fly	4	8
gall midge	20*	8*
evening primrose	7	14
onion	8	16
corn	10	20
grasshopper (female)	11	22
grasshopper (male)	10	21**
frog	13	26
sunflower	17	34
cat	19	38
human	23	46
plum	24	48
dog	39	78
sugar cane	40	80
goldfish	47	94

* In the fertilized egg of the gall midge, 32 chromosomes become nonfunctional leaving 8 functional chromosomes.
** The male grasshopper has only one sex chromosome.

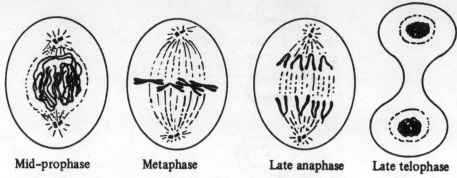

Mid–prophase   Metaphase   Late anaphase   Late telophase

Fig. 3.24 Stages of mitosis

MITOSIS

Mitosis (also known as karyokinesis) concerns the cell nucleus and its chromosomes. Before the onset of mitosis, the cell is in a stage known as *interphase*. During interphase, the chromosomes are exceptionally long and very thin, appearing as fine granules through the light microscope. It is during this stage that DNA molecules in the nucleus *replicate*. The result of replication is that each chromosome now has an exact copy of itself. When interphase comes to an end, the cell has enough nuclear material for two cells. The orderly process that divides the chromosomes equally between the two daughter cells is known as *mitosis*.

There are four stages of mitosis: prophase, metaphase, anaphase, and telophase. The significant events which mark each of these stages and interphase are outlined below and shown in Fig. 3.24.

Prophase
1. Chromosomes become shorter and thicker. Each chromosome consists of two *chromatids* attached at the centromere (Fig. 3.25).
2. The nuclear membrane begins to disintegrate.
3. Spindle fibers form extending from the centromeres to the poles.
4. The centrioles in animals cells, fungi, algae and some ferns replicate and a pair migrates toward each pole.
5. Chromosomes begin to move toward the equator of the cell.

Metaphase
1. The centrioles have migrated to the poles.
2. The chromosomes are lined up at the equator of the spindle.
3. Spindle fibers are attached to the centromeres connecting them to the poles of the spindle.
4. Both the nuclear membrane and the nucleolus have disappeared.

Anaphase
1. The centromeres split apart.
2. The chromatid pairs of each chromosome separate from each other. They move quickly in opposite directions, one toward each pole.

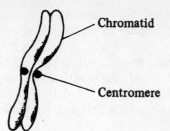

Chromatid

Centromere

Fig. 3.25 Chromatids hold together at centromere

Telophase
1. The recently separated chromosomes reach the poles. A pole is the place where the new nucleus of each daughter cell will be located.
2. The spindle fibers extending from the poles to the centromeres disappear. Those fibers that lie in the plane between the opposing rows of chromosomes remain for a longer time.
3. A nuclear membrane reforms around each bundle of chromosomes at the poles. At this time, all remnants of the spindle fibers have disappeared.
4. At the equator of animal cells, the cytoplasm turns inward, pinching the old cell into two new ones. In plant cells, a cell plate of rigid cellulose separates the two new cells.

Interphase
Two new daughter cells are formed identical in genetic material to the parent and to each other. In a relatively short span of time, the daughter cells will accumulate more cytoplasm and grow to the size of the original parent cell. The cell nucleus controls the biochemical activities of the cell during interphase. The only work that the "resting" cell is not doing is dividing.

CYTOKINESIS

Cytokinesis is the separation of the cytoplasm following nuclear division. At first there is a doubling of the molecules that make up the cytoplasm. Cell structures that are composed of protein subunits, such as microtubules, microfilaments and ribosomes, are synthesized from molecules within the cytoplasm. The membranous organelles such as the Golgi apparatus, lysosomes, vacuoles and vesicles are assembled by the membranes of the endoplasmic reticulum. The endoplasmic reticulum is restored and enlarged by self-assembling molecules. Chloroplasts and mitochondria replicate themselves from existing chloroplasts and mitochondria, respectively; it is theorized that these organelles have their own chromosome much like that in a prokaryotic cell. (Some scientists believe that the chloroplast and the mitochondrion were once independently living organisms that now live symbiotically in cells.)

The cytokinetic activities discussed above cannot be seen. The visible part of cytokinesis commences in late anaphase and reaches completion in telophase. The first sign of cytokinesis is the formation of a *cleavage furrow* which cuts across the equator of the cell through the spindle. In animal cells, microfilaments appear at the furrow which separates the daughter cells. In plants a cell plate separates the daughter cells. The cell plate is produced from a series of vesicles that are provided by the Golgi apparatus.

In some algae and fungi, mitosis without cytokinesis is common. When cytokinesis does not occur, cells with many nuclei but without separating membranes or walls result. Plant bodies made up of multi-nucleated cells without membrane or cell wall separations are described as being *coenocytic*.

MEIOSIS

Meiosis, or reduction division, is a kind of cell division that occurs in the primary sex cells leading to the formation of viable egg and sperm cells. The purpose of meiosis is to reduce the number of chromosomes to one half in each gamete so that upon fertilization (the fusing of sperm and egg nuclei) the species chromosome number is kept constant. If the gametes contained the full complement of chromosomes, on fertilization the number would be doubled. Instead, a gamete contains the haploid number of chromosomes, one of each type of chromosome.

Meiosis occurs in primary sex cells—oocytes in the female, spermatocytes in the male—which have the diploid chromosome number. The process includes two cell division events—Meiosis I and Meiosis II—coming one after the other (Fig. 3.26).

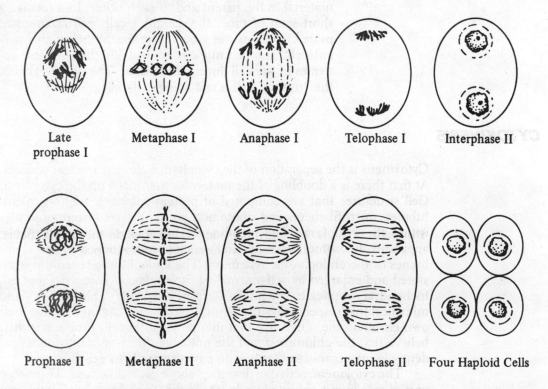

| Late prophase I | Metaphase I | Anaphase I | Telophase I | Interphase II |

| Prophase II | Metaphase II | Anaphase II | Telophase II | Four Haploid Cells |

Fig. 3.26 Stages of meiosis

Meiosis I

The stages of meiosis I result in the reduction of the number of chromosomes.

Prophase I
1. The chromosomes become shorter and thicker.
2. The nucleolus disappears.
3. Chromosomes pair with their homologues (mates) forming a group of four chromatids referred to as a *tetrad*.
4. The tetrads wrap around each other (synapse) and may exchange like parts.
5. The centrioles migrate and the spindle fibers appear.
6. The nuclear membrane disappears.

Metaphase I
1. The tetrads move as a unit to the equator.
2. The centromeres (kinetochores) of each of the homologous pairs of chromosomes become attached to spindle fibers extending from opposite poles.

Anaphase I
1. Each pair of double-stranded chromosomes (a set of sister chromatids) is pulled away from its homologue toward opposite poles.
2. The centromeres do not uncouple and the sister chromatids remain attached.

Telophase I
1. The chromosomes are double-stranded.
2. In some organisms the nuclear membrane reappears; in others, it does not and metaphase II starts immediately.

Interkinesis
1. The chromosomes disappear.
2. There is no replication of DNA.
3. Two haploid nuclei are present.
4. Interkinesis lasts for a very short time.

Meiosis II

This second stage separates the chromatids, terminating in four haploid cells.

Prophase II
1. The chromosomes reappear as do the spindle fibers.
2. The centrioles migrate to opposite poles.

Metaphase II
1. Spindles form.
2. The double-stranded chromosomes migrate to the equator. Their centromeres become attached to the spindle fibers.
3. The centromeres uncouple as they did in mitosis.

Anaphase II
The chromosomes pull apart to opposite poles.

Telophase II
1. Four haploid nuclei are formed. Each nucleus has one member of each pair of chromosomes that began the original meiosis.
2. The nuclear membrane reforms.
3. Cytokinesis comes to completion.

Mechanically, meiosis II is primarily a mitotic division. As in mitosis, the chromosomes do not synapse. Since the nucleus is now haploid, there are no homologous chromosome pairs. Each double-stranded chromosome moves to the equator independently, not being attached to the spindle with a homologue. Each haploid cell produced during meiosis I divides again during meiosis II, producing four new haploid cells.

CLONING

By definition a *clone* is a population of cells (or whole organisms) that has descended from an original parent cell, which was stimulated to reproduce by asexual means. This means that a clone is genetically identical to the original cell. Over the past 40 years, tissue culture experiments have been carried out with plants in which single, nonembryonic cells have been induced to develop along a pathway of a fertilized egg. The results with such species as carrot, African violet, Boston fern and Cape sundew indicate that a whole plant can be propagated from a single nonreproductive cell. Cloning techniques have been tried with animal cells with varying degrees of success.

Chronology of Famous Names in Biology

1665 **Robert Hooke** (English)—was the first to view the pores in cork through the microscope and applied the word "cell" to these box-like structures.

1702 **Anton von Leeuwenhoek** (Dutch)—described red blood cells and protists.

1828 **Robert Brown** (English)—discovered and described the nucleus in orchid plants.

1830 **Johann Evangelista Purkinje** (Czech)—described the nucleus of the hen's egg.

1835 **Felix Dujardin** (French)—described cell sap as the essential substance of life.

1838 **Matthias Schleiden** and **Theodor Schwann** (German)—formulated the cell theory: all living organisms are made of cells.

1839 **Johann Evangelista Purkinje** (Czech)—invented the term protoplasm.

1858 **Rudolf Virchow** (German)—founded the study of cellular pathology and discovered that new cells could arise only from existing cells.

1869 **Fredrick Miescher** (Swiss)—was the first to isolate nucleic acid which he did from pus cells.

1880 **Walther Flemming** (Austrian)—discovered and described mitosis in the cells of amphibious larvae. He called mitosis "the dance of the chromosomes."

1898   **Camillio Golgi** (Italian)—discovered the membranes known as the Golgi apparatus by staining cells with silver nitrate and osmium tetroxide.

1937   **Sir Hans A. Krebs** (English)—was the first to describe the citric acid cycle of cellular respiration. He was awarded the Nobel Prize for this work in 1953.

1938   **James Danielli** (American)—accurately described the phospholipid layers of the cell membrane.

1945   **Albert Claude** (Belgian) and **Keith Porter** (American)—discovered the endoplasmic reticulum.

1948   **Albert Lehninger** and **Eugene Kennedy** (American)—discovered that synthesis of ATP occurs in the mitochondria.

1949   **Christian de Duve** (Belgian)—discovered the lysosomes.

1951   **Maurice Wilkins** and **Rosalind Franklin** (English)—used X-ray diffraction to study DNA.

1952   **Hugh Huxley** (English)—elucidated the biochemical processes involved in muscle contraction.

1952   **Fritiof Sjostrand** (Swedish) and **George Palade** (American)—discovered cristae in the mitochondria.

1953   **Francis Crick** (English) and **James Watson** (American)—elucidated the double helix structure of DNA.

1954   **George Palade** (American)—discovered ribosomes. In 1956 he discovered the function of rough endoplasmic reticulum.

1954   **F. C. Steward** (American)—produced the first carrot clone from cells of the mature carrot root.

1956   **Philip Siekevitz** and **George Palade** (American)—isolated the ribosomes.

1956   **Joe Hin Tjio** (American) and **Albert Levan** (Swedish)—established the number of 46 for human chromosome count.

1960   **Hans Moor** (Swiss)—developed the freeze-fracture technique used to study the interior of membranes.

1962   **Earl W. Sutherland** (American)—discovered the role of cyclic AMP in the transport of certain hormones across the cell membranes. He was awarded the Nobel Prize for this work in 1971.

1962   **Marshall Nirenberg, Severo Ochoa,** and **Har Gobind Khorana** (American all)—deciphered the genetic code.

1962   **M. F. Perutz** and **J. C. Kendrew** (English)—awarded the Nobel Prize for discovering the structure of the muscle protein myoglobin.

1966   **S. J. Singer** (American)—proposed the "fluid mosaic" model of the cell membrane.

1967   **Edwin Taylor** (American)—discovered the role of microtubules in mitosis.

1976   **Har Gobind Khorana** (American)—produced an artificial gene.

1978   **Gunther Blobel** (American)—discovered a "signal sequence" for RNA.

1978   **Masayaso Nomura** (American)—took apart RNA proteins in subunits of bacterial ribosomes and put them back together again.

Words for Study

active transport	DNA	osmosis
active site	endoplasmic	oxidative
adenosine	reticulum	phosphorylation
diphosphate (ADP)	enzyme	pectin
adenosine	eukaryote	peptide bond
triphosphate (ATP)	flagella	phagocytosis
aerobic respiration	Golgi body	pinocytosis
amino acid	glycolysis	plasma membrane
anaerobic respiration	haploid number	plasmolysis
anaphase	inorganic	plastid
carbohydrate	interphase	polypeptide
carboxyl group	lipid	prophase
catalyst	lysosome	protein
cell	Krebs citric acid	protoplasm
cell membrane	cycle	pyruvic acid
cell plate	meiosis	replicate
cell theory	mesosome	resolving power
cellulose	metaphase	ribosome
centriole	microtubules	RNA
centromere	mitochondrion	semi-permeable
chloroplast	mitosis	stroma
chromatid	monosaccharide	substrate
chromosome	multicellular	system
cilia	nanometer	telophase
cleavage furrow	nucleoli	thylakoid
cristae	nucleic acid	tissue
cytokinesis	nucleus	turnover number
cytoplasm	organ	unicellular
diffusion	organelle	vacuole
diploid number	organic	vesicle

Questions for Review

PART A. **Completion.** Write in the word that correctly completes each statement.

1. Hereditary instructions are carried in the cell in the organelle named the ..1..

2. The general term given to the movement of materials into and out of cells is ..2..

3. The movement of water across a cell membrane is known as ..3..

4. Solid particles are ingested by cells through a process known as ..4..

5. The flow of circulating materials inside of a cell is directed by protein structures known as ..5..

6. The network of membranous canals that are spread through the cytoplasm are known collectively as the ..6..

7. Globular proteins are embedded in the ..7.. bilayer in the cell membrane.

8. Small storage areas in the cytoplasm of animal cells are known as ..8..

9. Proteins for export are synthesized by ..9.. that are attached to membranes of the endoplasmic reticulum.

10. The granular stage of the fine threads that are contained in the nucleus during interphase is known as ..10..

11. Cell walls of green plants are for the most part made of the compound ..11..

12. Plant cells are colored green by the pigment named ..12..

13. Protists use for locomotion fine cytoplasmic hairs known as flagella and ..13..

14. The type of tissue that covers and lines organs is ..14.. tissue.

15. Cells specialized for the carrying of nervous impulses are nerve cells, also known as ..15..

16. Amino acids are the building molecules of the nutrient molecules ..16..

17. Types of compounds that control chemical reactions are organic catalysts, also known as ..17..

18. The initials ATP are an abbreviation for ..18..

19. The symbol 2N refers to the ..19.. number of chromosomes.

20. Meiosis takes place in egg and sperm cells which are known collectively as ..20..

PART B. **Multiple Choice.** Circle the *letter* of the item that correctly completes each statement.

1. Multicellular organisms are produced from
   (a) one cell   (c) three cells
   (b) two cells   (d) four cells

2. The fine structures of the cell can be seen with the
   (a) naked eye
   (b) light microscope
   (c) phase contrast microscope
   (d) electron microscope

3. The cell's energy is used in the process of
   (a) passive transport
   (b) active transport
   (c) diffusion
   (d) osmosis

4. Because the cell membrane is highly selective in regard to the materials that can cross its boundary, it is described as being
   (a) semi-porous
   (b) semi-permeable
   (c) semi-fluid
   (d) semi-solid

5. The shrinking of cytoplasm due to the loss of water molecules is known as
   (a) evaporation
   (b) osmosis
   (c) diffusion
   (d) plasmolysis

6. Molecules carried to the cell surface by vesicles are forced out of the cell by the process of
   (a) exocytosis
   (b) endocytosis
   (c) facilitated transport
   (d) adsorption

7. The most numerous organelles in the cell are the
   (a) mitochondria
   (b) lysosomes
   (c) ribosomes
   (d) microtubules

8. Fluid substances for export outside of the cell are stored temporarily in the membranes of the
   (a) ribosomes
   (b) lysosomes
   (c) endoplasmic reticulum
   (d) Golgi apparatus

9. RNA is synthesized and temporarily stored in the
   (a) ribosomes
   (b) mitochondria
   (c) nucleoli
   (d) lysosomes

10. The internal membranes of the mitochondria are known collectively as
    (a) grana
    (b) thylakoids
    (c) mesosoma
    (d) cristae

11. Cell sap is stored in areas of plant cells known as
    (a) lysosomes
    (b) centrosomes
    (c) vesicles
    (d) vacuoles

12. Thylakoids are the same as
    (a) cristae
    (b) nucleoli
    (c) grana
    (d) stroma

13. The type of tissue that is specialized to respond to stimuli transmitted by motor nerve cells is
    (a) blood
    (b) muscle
    (c) connective
    (d) bone

14. A nonliving matrix is most correctly associated with the type of tissue classified as
    (a) epithelial
    (b) muscle
    (c) connective
    (d) nerve

15. A type of plant tissue specialized for conducting water and its dissolved materials is
(a) vascular
(b) epidermal
(c) fundamental
(d) collenchyma

16. Substrate molecules are acted upon by
(a) enzymes
(b) hormones
(c) carbohydrates
(d) lipids

17. Enzymes are synthesized from organic molecules known as
(a) minerals
(b) carbohydrates
(c) proteins
(d) polysaccharides

18. Anaerobic respiration is correctly known as
(a) hydrolysis
(b) osmosis
(c) glycolysis
(d) plasmolysis

19. The aerobic phase of respiration takes place inside of the
(a) cytoplasm
(b) nucleus
(c) lysosome
(d) mitochondrion

20. The circular chromosome of a prokaryotic cell attaches itself to a site on the cell membrane named the
(a) centrosome
(b) kinetosome
(c) mesosome
(d) centriole

PART C. **Modified True-False.** If a statement is correct, write "true" for your answer. If a statement is incorrect, change the underlined word to a word that will make the statement true.

1. Scientists believe that cell shape seems to be related to its <u>age</u>.

2. Groups of similar cells that work together to do a particular job are designated as <u>systems</u>.

3. Large organisms have <u>fewer</u> cells than small organisms.

4. Hereditary instructions are carried in molecules of <u>CBA</u>.

5. The parts of a cell are known as <u>organs</u>.

6. Passage of materials into and out of the cell is controlled by the <u>cell wall</u>.

7. Another name for "cell drinking" is <u>imbibing</u>.

8. All of the material that is outside of the nucleus and inside of the cell membrane is called <u>protoplasm</u>.

9. Digestive enzymes and hormones are examples of <u>lipid</u> molecules that are exported from cells.

10. Aside from the nucleus, the <u>centriole</u> is the largest organelle in the cell.

11. Areas in plant cells that store starch and pigment molecules are given the general name of <u>plastics</u>.

12. The food-making process of green plants is known as <u>phototropism</u>.

13. The names Schleiden and Schwann are correctly associated with the chromosome theory.

14. A type of white blood cell specialized to counteract disease organisms that enter the body is the erythrocyte.

15. Another name for involuntary muscle is striated muscle.

16. Tracheids and vessels are best associated with phloem tissues.

17. The Krebs' citric acid cycle is the anaerobic phase of respiration.

18. As a result of mitosis, the daughter cells receive the haploid number of chromosomes.

19. The number of chromosomes in a human skin cell is 23.

20. Chromosome replication takes place during prophase.

Answers to Questions for Review

PART A

1. nucleus
2. transport
3. osmosis
4. phagocytosis
5. microtubules
6. endoplasmic reticulum
7. phospholipid
8. vesicles
9. ribosomes
10. chromatin
11. cellulose
12. chlorophyll
13. cilia
14. epithelial
15. neurons
16. proteins
17. enzymes
18. adenosine triphosphate
19. diploid
20. gametes or sex cells

PART B

1. a
2. d
3. b
4. b
5. d
6. a
7. c
8. d
9. c
10. d
11. d
12. c
13. b
14. c
15. a
16. a
17. c
18. c
19. d
20. c

PART C

1. function
2. tissues
3. more
4. DNA
5. organelles
6. cell membrane
7. pinocytosis
8. cytoplasm
9. protein
10. mitochondrion
11. plastids
12. photosynthesis
13. cell
14. leucocyte
15. smooth
16. xylem
17. aerobic
18. diploid
19. 46
20. interphase

CHAPTER 4

BACTERIA, CYANOBACTERIA AND VIRUSES

Bacteria and cyanobacteria (blue-green algae) are classified in the kingdom Monera. Virus particles are not living cells and do not belong in the classification system of living things. However, viruses play an important part in our lives. They are useful as experimental materials in gene studies and significant as causes of diseases in humans, animals and plants. For lack of a better place to include virus study, the author has placed it here where these particles can be easily compared with living cells.

The Kingdom Monera

The monerans are prokaryotic cells. (Some authors use the spelling procaryotic). They lack a nuclear membrane, mitochrondria, an endoplasmic reticulum, Golgi apparatus and lysosomes. As a matter of fact, the prokaryotes do not have any membrane-bound organelles in the cytoplasm. The Monera are unicellular organisms usually invisible to the naked eye. Most monerans live as independent cells, although some may occur in filaments (chains of cells) or colonies of cells held together by a gelatinous coat.

GENERAL CHARACTERISTICS OF THE PROKARYOTES

Biologists agree that three basic differences in structure separate prokaryotic cells from the eukaryotes. First, the chemical composition of the prokaryotic cell wall differs significantly from that of the eukaryotic cell wall. Secondly, the prokaryotes lack an organized nucleus and are devoid of membrane-bound organelles in the cytoplasm. Thirdly, the genetic material in prokaryotes is structured into a single chromosome in which the DNA is intermingled with proteins that differ from those in eukaryotic nuclei. Other differences, as well, separate the nonnucleated cells from those cells with a nucleus. Most prokaryotic cells are quite small, ranging in size from 0.5 to 10 micrometers. The average size of the eukaryotic cell is much larger, although some nucleated cells measure only 7 micrometers in diameter (Fig. 4.1).

Cell Membrane

The cell membrane of the prokaryotic cell is very much like that of the eukaryotic cell except that it lacks cholesterol and other steroids. In some prokaryotes, the surface area of the cell membrane is greatly increased by *convolutions* (folds and loops). The convoluted cell membranes have incorporated in their structure electron transport systems and enzymes necessary for the chemical events taking place during respiration. The circular DNA molecule is attached to a site on the cell membrane called the *mesosome*.

Cell Wall

Present in most prokaryotic cells, the cell wall is an important structure providing shape to the cell and protecting it against osmotic abuse. (However, *mycoplasmas*, the smallest living cells, do not have a cell wall.) Ranging from 5 to 80 nanometers in diameter, the walls of prokaryotic cells are composed of a complex polysaccharide substance known as *murein*. Muramic acid, the main constituent of murein, never appears in the eukaryotes. The antibiotic penicillin is toxic to multiplying bacteria because it inhibits the production of murein and thus prevents the formation of the cell wall.

Cytoplasm

The cytoplasm of the prokaryotes does not have the complex fine structure characteristic of the eukaryotes. Bacterial cytoplasm appears to be highly granulated due to the presence of a large number of ribosomes which are smaller in size than those found in nucleated cells. The ribosomes—the only cellular organelles in most prokaryotes—function just the same as their larger counterparts in the eukaryotic cells by being the sites of protein synthesis. A peripheral membrane system is present in cells of the cyanobacteria; these membranes house chlorophyll and accessory pigments.

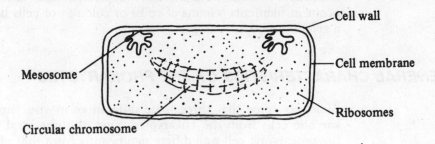

Fig. 4.1 A prokaryotic cell

Genetic Material

Prokaryotic cells contain both types of nucleic acid: RNA and DNA. Just as in eukaryotic cells, mRNA is used in protein synthesis. The DNA of the prokaryotes is not like eukaryotic DNA in that the former is interspersed with different kinds of proteins. Prior to cell division, the circular DNA *replicates*, producing another DNA molecule identical to itself.

MAJOR GROUPS OF MONERANS

Organisms grouped in the kingdom Monera are somewhat biological puzzles, and there are divergent opinions about the best method of classifying these microorganisms. For example, bacteria were at one time classified as one-celled plants. However, in general, the kingdom Monera now has assigned to it three phyla: the Schizomycetes (bacteria), Cyanophyta (cyanobacteria) and the Prochloropyta.

Schizomycetes—Bacteria

Bacteria are the smallest living organisms. They range in length from 0.2 to 7 micrometers; in diameter from 0.2 to 2 micrometers. As you recall, the unit used to measure bacteria is the micrometer and is equivalent to $\frac{1}{1000}$ of a millimeter. The smallest cells known are the mycoplasmas which have only one-half the DNA of other bacterial cells. The mycoplasmas are bacteria that live only as parasites on or in the bodies of plants and animals.

Despite their small size bacteria are true cells: they provide their own genetic material (DNA and RNA) and the necessary cytoplasm for their own reproduction; they have multienzyme systems to control biochemical activities necessary for the life of the cell; and they build their own ATP molecules and use the stored energy to synthesize other organic compounds.

Classes of Bacteria

The phylum Schizomycetes includes five classes of bacteria. Each class consists of numerous species. Many biologists refer to "divisions" or groups of bacteria, because the concept of species is difficult to apply to these organisms. By accepted definition, a species is a group of organisms so closely related that they can mate and produce viable offspring. Since bacteria usually reproduce asexually and since their evolutionary history is difficult to substantiate, many species of bacteria do not easily meet the specifications of the definition. However, we shall continue to use the word "species" to show close structure and functional relationship between bacterial organisms.

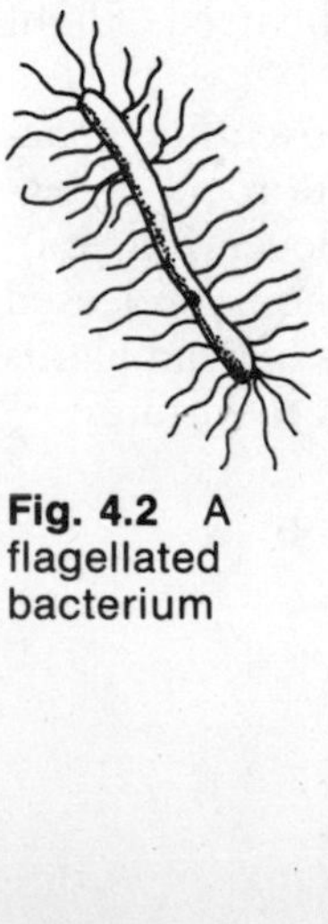

Fig. 4.2 A flagellated bacterium

Eubacteria—True Bacteria   This group, referred to as the true bacteria, represents a large number of species. All of these bacteria have thick and rigid cell walls. Some of the species are *nonmotile* (nonmoving), while others are *motile*, using flagella (Fig. 4.2) or a sling motion to move from place to place.

Species belonging to the eubacteria are identified by their shapes. The rod-shaped bacteria are known as *bacilli* (*bacillus*, sing.), the round bac-

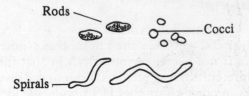

Fig. 4.3 Three shapes of bacteria

teria as the *cocci* (*coccus*, sing.), and the spiral-shaped as *spirillae* (*spirillum*, sing.) (Fig. 4.3). Some species typically remain attached. The diplococci occur in pairs, the streptococci in chains and the staphylococci in clusters.

Reproduction in the eubacteria takes place by *binary fission*. There is no mitosis as in the eukaryotic cells. Replication of the circular chromosome is followed by equal splitting of the cytoplasm. The reproductive potential of bacteria is enormous because cell division is possible every 20 minutes. In this short span of time, a parent cell divides into two identical daughter cells. It has been estimated if the rate of reproduction were not slowed by environmental changes, one bacterium at the end of 6 hours could give rise to 500,000 descendants.

Eubacteria are able to survive unfavorable environmental conditions in the rather unique way of forming *endospores*. The spore is a vegetative cell containing DNA and little else. It becomes surrounded by a thick and almost indestructible cell wall secreted by the cytoplasm. In this condition, the endospore is able to withstand boiling, freezing, drying, treatment with disinfectants and other extremes of environment.

Capsule formation occurs in certain disease-producing species of eubacteria. The cytoplasm of the bacterium secretes a mucoid coat of polysaccharide materials which surrounds the cell wall. The capsule is resistant to phagocytosis (engulfing by white blood cells) and may even, itself, release toxins into the tissues of the host.

Most of the eubacteria are *heterotrophs* that have to depend on sources outside of their own bodies for organic food materials. Many species obtain food by absorbing nutrients from dead organic matter; organisms that feed in this way are known as *saprophytes*, or, in new terminology, *saprobes*. Bacteria that live on or inside of the bodies of animals or plants and cause disease are *parasites*.

A number of species in the eubacteria are *autotrophs* and are able to change inorganic materials into organic compounds. Among these are the *photosynthetic bacteria*, which, like green plants, use light energy to produce food, and the *chemosynthetic bacteria*, which are capable of oxidizing the inorganic compounds of ammonia, nitrites, sulfur, or hydrogen gas into high-energy organic compounds without the need of light energy.

Nitrifying bacteria are excellent examples of the chemosynthetic bacteria. These bacteria live in little *nodules* (bumps) on the roots of leguminous plants such as peas, beans, peanuts, alfalfa and clover (Fig. 4.4). In two separate steps, the nitrifying bacteria convert ammonia, released into the soil by the breakdown of proteins from the bodies of dead plants and animals, to nitrites and then they change the nitrites to nitrates:

(1) $2NH_4^+ + 3O_2 \rightarrow 2NO_2^- + 4H^+ + \text{energy}$

(2) $2NO_2^- + O_2 \rightarrow 2NO_3^- + \text{energy}$

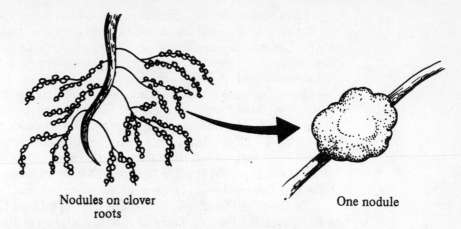

Nodules on clover
roots

One nodule

Fig. 4.4 Nodules on the roots of leguminous plants

Rod-shaped bacteria belonging to the genus *Nitrobacter* are examples of nitrifying bacteria. Bacteria of decay, responsible for the breakdown of proteins in dead organic matter, belong to the genera *Pseudomonas* and *Thiobacillus*. Species belonging to the nitrifying and denitrifying genera aid in the recycling of nitrogen (Fig. 4.5).

Sulfur bacteria are also chemosynthetic bacteria. Sulfur, an element necessary for life, is present in soil, free and combined, in inorganic and organic compounds. The decomposition of rock, the breakdown of organic products, and rainwater are the major sources of soil sulfur. Living in the soil, the "sulfur" bacteria oxidize free sulfur in sulfates, producing energy for themselves and providing compounds of sulfur that can be used by plants.

$$2S + 3O_2 + 2H_2O \rightarrow 2SO_2 + 4H^+ + \text{energy}$$

Photosynthetic bacteria (Endothiobacteria) use pigment systems which are contained in the cell membrane to capture light energy and use it in the synthesis of organic molecules. Unlike green plants, however, photosynthetic bacteria do not have the pigment chlorophyll *a* and they do not produce molecular oxygen. In fact, most photosynthetic bacteria do not need or use oxygen.

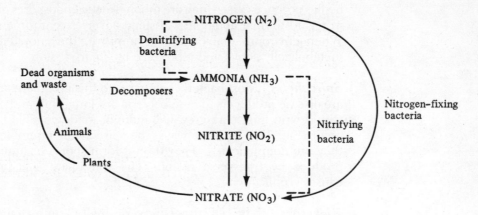

Fig. 4.5   Nitrogen cycle

Most species of the eubacteria are *aerobic*, using molecular oxygen in the process of breaking down carbohydrates to carbon dioxide and water. *Obligate aerobes* are those organisms that can live only in an environment that provides free or atmospheric oxygen. An example of an obligate aerobe is *Bacillus subtilis*.

Some bacteria are *obligate anaerobes* and derive their energy by fermentation. These organisms cannot live in an environment of free oxygen. Usually, the obligate anaerobes are disease producers; included in this group are *Clostridium tetani*, the causative organism of tetanus, and *Clostridium botulinum*, the bacterium that induces food poisoning. Still other bacteria are *facultative anaerobes*. These are basically aerobic bacteria, but they can live and grow in an environment that lacks free oxygen.

In 1884 the Danish microbiologist Hans Christian Grams developed a technique for staining bacteria that is useful in distinguishing groups of organisms. The procedure of *Gram staining* identifies cells based on their ability to absorb certain dyes. Bacterial cells walls that are made of peptidoglycans (amino sugars) absorb the purple dye gentian violet with such tenacity that washing the cells with alcohol does not remove the dye. Such purple-staining cells are designated as *Gram-positive*. Cell walls composed of lipopolysaccharides do not hold the purple dye and are easily bleached with alcohol. These cells stain red with dyes such as safranin or carbol fuchsin. Cells that take up the red counterstain are known as *Gram-negative*.

Gram staining is used as one means by which bacteria are identified and classified. As a matter of fact, the composition of the cell wall affects the behavior of bacteria. Gram-positive bacteria are more susceptible to the effects of antibiotics than are Gram-negative organisms. Gram-positive organisms are also more susceptible to the lysing effects of the enzyme lysozyme, which is found in human secretions such as tears, saliva, and nasal discharges.

Myxobacteria—Slime Bacteria The Myxobacteria are also known as the slime bacteria. The cells develop as a colony and move together in a flowing mass of slime resembling the movements of amoebae. The individual cells are long, thin rods. Myxobacteria live in the soil and obtain nutrients by absorbing dead organic matter (saprobes). At times the slime mass, or *pseudoplasmodium*, forms reproductive structures called *fruiting bodies* which look like small mushrooms. The fruiting bodies produce *cysts* which are much like the endospores of other bacteria in regard to their ability to withstand adverse environmental conditions. When environmental conditions improve, the encysted cells become metabolically active and reproduce by binary fission.

Spirochetes Most bacteria belonging to this group are anaerobic; many are disease producers. These bacteria are long, thin and curved, moving with a wriggling, corkscrew-like motion, made possible by an axial filament (Fig. 4.6). In some ways, the spirochetes resemble protozoa, but they are nonnucleated. They do not form spores or branches. They reproduce by transverse fission. The spirochete *Treponema pallidum* causes syphilis.

Rickettsiae The rickettsiae are very small Gram-negative intracellular parasites (Fig. 4.7). They were first described in 1909 by Harold Taylor

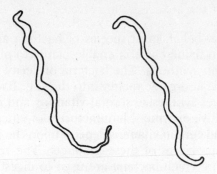

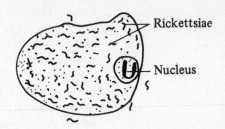

Fig. 4.6 Spirochetes are highly motile but they lack polarity

Fig. 4.7 Rickettsiae in tissue cells

Ricketts, who found them in the blood of patients suffering from Rocky Mountain spotted fever. The rickettsiae are nonmotile, nonspore forming, nonencapsulated organisms. In length, these organisms range from 0.3 to 1 micrometers. They live in the cells of ticks and mites and are transmitted to humans through insect bites. The rickettsiae are responsible for several febrile (fever-producing) diseases in humans, such as typhus fever, trench fever and Q fever.

Actinomycetes The Actinomycetes (also known as the Actinomycota) have characteristics of both the eubacteria and the fungi. Like the latter, their body structure takes the form of branching multicellular filaments. They live in soil where they obtain nutrition saprophytically. Some species are anaerobes. These organisms are mostly nonmotile and break down waxes and lipids in dead plants and animals. Some species cause diseases in animals and humans. *Actinomyces bovis* causes "lumpy" jaw in cattle. *Mycobacterium tuberculosis* causes tuberculosis in humans and *Mycobacterium leprae* causes leprosy. Some other actinomycetes have important medical and commercial value—primarily as sources of *antibiotics*. Antibiotics in common use that are produced by the actinomycetes are tetracycline, erythromycin, neomycin, nystatin and chloramphenicol.

Bacteria and Disease

As we have already mentioned, some species of bacteria cause disease in humans, animals and plants; these bacteria are termed *pathogenic*. A disease is any condition that interrupts the normal functioning of body cells preventing completion of a particular biochemical task. When pathogenic bacteria invade body tissues, they introduce a particular set of circumstances that change the environment of the cells.

There are three significant ways in which the cellular environment can be altered resulting in a disease condition. One way is by sheer numbers of organisms: enormous numbers of bacteria will affect adversely the functioning of cells. For example, *Escherichia coli*, Gram-negative rods that normally live in the human intestine will cause disease if present in increased numbers. Another way is by the destruction of cells and tissues. A third change brought about by pathogenic bacteria is directly related to their production of *toxins*. Toxins are poisonous substances that inhibit the metabolic activities of the host cells.

Helpful Bacteria

Contrary to popular belief, most species of bacteria are helpful to humans. The nitrogen-fixing bacteria enable plants to obtain the nitrates necessary for protein synthesis. The bacteria of decay release ammonia and nitrates from dead organic matter into the soil. Bacteria that live in the human intestines synthesize several vitamins and contribute to the synthesis of a digestive enzyme. Manufacturers of vinegar, acetone, butanol, lactic acid and certain vitamins depend upon the action of bacteria in the production processes of these products. The retting of flax and hemp is a process in which bacteria are used to digest the pectin compounds that hold together the cellulose fibers. Once these fibers are free they can be used to make linen, textiles and rope. Bacteria are useful in the preparation of skins for leather and in the curing of tobacco. Manufacturers of dairy products use bacteria to ripen cheese and to improve the flavor of special items such as Swiss cheese. Farmers depend on bacteria in their fermenting of silage that is used for cattle feed. The pharmaceutical industry produces antibiotics such as aureomycin, terramycin and streptomycin from bacteria.

Cyanophyta

A controversial phylum in the kingdom Monera is the Cyanophyta. Some modern biologists classify these one-celled photosynthetic organisms as blue-green algae and thus the name Cyanophyta. Other biologists regard these organisms as bacteria and call them *cyanobacteria*. Still others avoid the issue and refer to this group simply as the blue-greens.

In structure, the cyanophytes are prokaryotes. Like the bacteria the cytoplasm of the blue-greens is crowded with ribosomes. In addition, the cytoplasm of most of these organisms stores particles of protein materials and special carbohydrate molecules known as *polyglucans*. The cells of the blue-greens are larger than bacteria but smaller than eukaryotic cells. The cell wall is composed of murein, as in the bacteria, but is surrounded by a gelatinous material composed of pectin and glycoproteins. The cyanobacteria have no visible means of locomotion (no flagella), but some species move with a gliding motion as do the myxobacteria.

The blue-greens derive the name from the pigment molecules that are contained in flattened membranous vesicles called *thylakoids*. Several types of pigments have been isolated from the cyanophyta. As in the higher plants, chlorophyll *a* is the pigment used for photosynthesis. Present also in these organisms is the blue pigment *phycocyanin*, the red pigment *phycoerythrin*, and several types of carotenoids (yellow pigments). Not all of the cyanobacteria are blue-green. Some are black, brown, yellow, red and bright green. At times the Red Sea takes on a reddish hue due to a species of cyanobacteria that inhabits these waters in great numbers and contains large amounts of phycoerythrin.

The cyanophyta may live singly or occur in filaments or colonies. Although the cells of some species live in close association with one another, they usually carry on independent life functions. Interestingly enough, cytoplasmic bridges called *plasmadesmata* join the cytoplasm of some of the cells of the filamentous groups. At certain times the plasmadesmata may serve as the passageways for special materials.

Some of the filamentous blue-greens have the ability to fix atmospheric nitrogen. Under anaerobic conditions, special cells called *heter-*

ocysts produce the enzyme *nitrogenase*, which has the ability to convert atmospheric nitrogen into usable nitrates.

Fresh water is the habitat of most of the cyanophyta, but a few species are marine, and some species live in tropical soils where the nitrogen content is poor. Other habitats include microfissures in desert rocks where light penetrates and a small amount of water is trapped. Cyanophytes can also be found on tree bark, on damp rocks and on flower pots. They reproduce by binary fission or by fragmentation of filaments.

The Cyanophyta are an important part of every environment. In the process of photosynthesis they produce oxygen as a by-product which helps to replenish the atmosphere. They also produce food for invertebrate and vertebrate species that feed on *phytoplankton*, the floating plants that live on the surfaces of oceans and lakes.

Cyanophyta can also be harmful to humans and fish in their role as water pollutants. The rapid overpopulation of blue-green algae such as *Oscillatoria* coats lakes with a slimy, smelly mass that is toxic to fish. The "blooms," aided by excessive nitrate compounds in lakes, have contributed to a serious decline in the freshwater fish populations in some areas.

Prochlorophyta

The Prochlorophyta were discovered in 1976 by R. A. Levin. These cells are prokaryotic and seem to be an evolutionary link between the Cyanophyta and the eukaryotic algae. Their pigment system seems to resemble that of the green algae and higher green plants. The only known representatives are those that live in association with a group of marine invertebrates known as the tunicates.

Viruses _____

As we said earlier, viruses are not living. They are not cells and they do not exhibit the characteristics of life as do cells. Viruses can reproduce only within living cells.

STRUCTURE OF VIRUSES

The virus particle is known as a *virion*. It merely consists of a protein coat and a nucleic acid core. The protein coat, called a *capsid*, may be shaped like a rod or may be polyhedral or may have a tail with extending fibers (Fig. 4.8). In some viruses, including those that cause influenza, a cytoplasmic membrane surrounds the protein coat. This surrounding *envelope* may come from the plasma membrane of the host cell or may be synthesized by the host's cytoplasm. However, virologists have found that the envelope contains proteins that are virus-specific.

Viral nucleic acid may be a single molecule consisting of as few as five genes or may have as many as several hundred. Viral nucleic acid may be single or it may be double-stranded; it may be circular or linear. Some virus nucleic acid is made of DNA (deoxyribonucleic acid); others have only an RNA (ribonucleic acid) core. Viruses never contain both DNA and RNA.

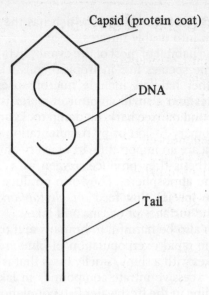

Fig. 4.8 Structure of a virus

REPRODUCTION OF VIRUSES

A general description of the way in which viruses reproduce follows. A virus infects (enters) a host cell. The nucleic acid of the virus captures the nucleic acid of the host cell and uses it to make more virus nucleic acid and virus proteins. New virus particles, complete with the nucleic acid core and the protein coat, are assembled inside of the host cell until no more can fit. The host then bursts, releasing the newly assembled virions. This generalized pattern of virus reproduction varies with different viruses that typically infect bacteria, plant, or animal cells. A virus that infects a bacterial cell is known as a *bacteriophage*, or simply as a *phage*.

IMPORTANCE OF VIRUSES

Viruses are notorious for their disease-producing potential in animals and plants. In human beings, viruses cause such diseases as venereal herpes, fever blisters, measles, the common cold, smallpox, rabies, chicken pox, yellow fever, influenza, viral pneumonia, encephalitis, infectious mononucleosis, and several types of hepatitis. It is believed that some cancers are caused by viruses. Virus diseases of plants include tobacco mosaic disease, tobacco necrosis and rice dwarf.

VIROIDS

In the 1960's, it was discovered that several plant diseases are caused by *viroids*, particles smaller than viruses. The viroid contains a short RNA chain and lacks the protective protein coat. New viroids are produced in the nucleus of the host cell. Diseases such as the stunting of chrysanthemums, spindle tuber of potatoes and disease of citrus trees are caused by viroids.

Chronology of Famous Names in Biology

1684	**Anton von Leeuwenhoek** (Dutch)—was the first person to see bacteria.
1796	**Edward Jenner** (English)—demonstrated the use of cowpox as an immunizing vaccine against smallpox.
1850	**Ignaz Semmelweis** (German)—discovered that attending physicians carried the organisms that cause childbed fever from patient to patient on their unwashed hands.
1861	**Louis Pasteur** (French)—made major contributions to the understanding of bacteria; helped to prove the germ theory of disease; discovered a method for the prevention of rabies.
1867	**Joseph Lister** (English)—developed antiseptic principles in the practice of surgery.
1884	**Robert Koch** (German)—discovered the bacteria that cause tuberculosis and cholera. He formulated Koch's postulates, the steps to determine if a bacterium causes a disease.
1884	**Hans Christian Gram** (Danish)—developed differential staining techniques for bacteria which have become known as the Gram stain.
1929	**Alexander Fleming** (English)—discovered lysozyme in human secretions.
1935	**Wendell Stanley** (American)—isolated the tobacco mosaic virus.
1946	**Joshua Lederberg** and **Edward Tatum** (American)—were the first to demonstrate genetic recombination in bacteria.

Words for Study

aerobic	eubacteria	obligate aerobe
autotroph	facultative	obligate anaerobe
bacillus	anaerobe	parasite
bacteriophage	fruiting body	pathogenic
binary fission	Gram staining	phage
capsid	heterocyst	photosynthetic bacteria
chemosynthetic	heterotroph	plasmadesmata
bacteria	mesosome	spirillum
coccus	murein	spirochete
convolution	monera	saprobe
cyanobacteria	motile	toxin
cyanophyta	mycoplasma	virion
endospore	nodule	viroid

Questions for Review

PART A.   Completion. Write in the word that correctly completes each statement.

1. The monerans are cells that lack ..1.. organelles. (2 words)
2. The cell membrane of the prokaryotes lacks the steroid ..2..
3. The DNA molecule is attached to a place on the cell membrane of bacteria called a ..3..
4. Mycoplasmas, unlike other prokaryotic cells, do not have a ..4.. (2 words)
5. Polysaccharide necessary for cell wall formation in bacteria is ..5..
6. The only cellular organelle in bacteria is the ..6..
7. A chromosome makes an identical copy of itself in a process called ..7..
8. The phylum Schizomycetes includes the ..8..
9. A group of interbreeding organisms is called a ..9..
10. The ..10.. are referred to as the true bacteria.
11. Bacteria reproduce by the method known as ..11..
12. In newer terminology, a saprophyte is referred to as a ..12..
13. Rod-shaped bacteria are known as ..13..
14. *Nitrobacter* are examples of ..14.. bacteria.
15. Bacteria that require molecular oxygen for respiration are known as ..15..
16. Bacteria that are indifferent to oxygen are called ..16..
17. Gram-positive bacteria absorb a ..17.. colored dye.
18. Toxins inhibit the ..18.. activities of cells.
19. True bacteria can survive unfavorable environmental conditions by forming ..19..
20. The ..20.. are evolutionary links between prokaryotic and eukaryotic cells.
21. In some cyanophyta cells known as ..21.. can fix atmospheric nitrogen.
22. The habitat of most blue-greens is ..22.. water.
23. A virus particle is known as a ..23..
24. A virus that infects bacteria is called a ..24..
25. Floating green plants that inhabit surfaces of lakes and oceans are known collectively as ..25..

PART B.   Multiple Choice. Circle the letter of the item that correctly completes each statement.

1. The monerans are best described as
   (a) eukaryotic cells   (c) prokaryotic cells
   (b) photosynthetic cells   (d) filamentous cells

2. A chain of cells is known as a
   - (a) filament
   - (b) mass
   - (c) chord
   - (d) convolution

3. The DNA of prokaryotes is structured into
   - (a) scattered chromatin granules
   - (b) histone chains
   - (c) several nucleoids
   - (d) a single chromosome

4. Muramic acid is isolated from the bacterial structure known as
   - (a) mesosome
   - (b) cell membrane
   - (c) cell wall
   - (d) chromosome

5. The function of the ribosomes is to
   - (a) synthesize cell walls
   - (b) carry hereditary traits
   - (c) direct replication of the mesosome
   - (d) translate mRNA into protein

6. Prokaryotic cells contain
   - (a) DNA only
   - (b) RNA only
   - (c) both DNA and RNA
   - (d) no nucleic acids

7. Of the following, the incorrect statement concerning bacteria is that they
   - (a) build ATP molecules
   - (b) lack multienzyme systems
   - (c) lack Golgi bodies
   - (d) assemble proteins

8. Survival of the eubacteria is increased by the ability to
   - (a) replicate DNA
   - (b) synthesize proteins
   - (c) generate cells walls
   - (d) form endospores

9. Encapsulated bacteria resist
   - (a) reproduction
   - (b) endospore formation
   - (c) phagocytosis
   - (d) DNA replication

10. Nitrifying bacteria
   - (a) convert ammonia into nitrates
   - (b) convert ammonia into nitrogen gas
   - (c) release ammonia from decaying bodies
   - (d) synthesize legumes

11. The bacteria that live as parasites in the cells of ticks and mites are
   - (a) spirochetes
   - (b) actinomycetes
   - (c) mycoplasmas
   - (d) rickettsiae

17. Lysozyme is a (an)
   - (a) human secretion
   - (b) small vacuole
   - (c) enzyme
   - (d) H carrier molecule

18. Fruiting bodies are best associated with
   - (a) spirochetes
   - (b) Eubacteria
   - (c) Prochlorophyta
   - (d) Myxobacteria

19. The organism that causes syphilis is a
   - (a) spirochete
   - (b) eubacteria
   - (c) prochlorophyte
   - (d) myxobacterium

20. Streptomyces are best classified as
   - (a) molds
   - (b) bacteria
   - (c) fungi
   - (d) protists

21. Plasmadesmata are
    (a) shrinking cell walls
    (b) oversized vacuoles
    (c) cytoplasmic bridges between cells
    (d) several attached cell membranes

22. Heterocysts are most closely associated with
    (a) nitrogen fixation
    (b) cell protection
    (c) water storage
    (d) binary fission

23. Actinomycetes are a type of
    (a) virus
    (b) rock
    (c) fungus
    (d) bacteria

24. The protein coat of a virus is known as a
    (a) desmid
    (b) plastid
    (c) capsid
    (d) capsule

25. A virus
    (a) can reproduce itself independently
    (b) can reproduce only within living cells
    (c) does not contain either DNA or RNA
    (d) can be considered a type of cell

PART C.   Modified True-False. If a statement is true, write "true" for your answer. If a statement is incorrect, change the underlined word to one that will make the statement true.

1. At one time bacteria were classified as one-celled <u>plants</u>.

2. Usually, Monera are <u>visible</u> to the naked eye.

3. The smallest living cells are <u>bacteria</u>.

4. In addition to the circular chromosome the only other organelle found in bacteria is the <u>Golgi body</u>.

5. A micrometer is <u>$\frac{1}{100}$</u> of a millimeter.

6. Mycoplasmas have <u>twice</u> as much DNA as the true bacteria.

7. <u>Photosynthetic</u> bacteria do not need light energy to synthesize food molecules.

8. During favorable conditions, bacteria can reproduce every <u>2 hours</u>.

9. Disease-producing bacteria may be surrounded by a polysaccharide <u>cell wall</u>.

10. Small bumps on roots of clover plants that house bacteria are known as <u>tumors</u>.

11. Round bacteria are called <u>spirillae</u>.

12. *Escherichia coli* live normally in the human <u>heart</u>.

13. When oxygen is not available, facultative anaerobes obtain energy by way of <u>the Krebs cycle</u>.

14. Obligate anaerobes must live in an environment free of <u>carbon dioxide</u>.

15. Bacterial cells <u>can</u> build their own ATP molecules.

16. <u>Gram-positive</u> bacteria are more susceptible to the effects of antibiotics.

17. Most species of bacteria are <u>harmful</u>.

18. Retting is a process in which <u>viruses</u> are used to digest the pectin in
flax plants.

19. Hemp fiber is used to make <u>linen</u>.

20. Polyglucans are molecules of <u>protein</u>.

21. The enzyme nitrogenase is produced under <u>aerobic</u> conditions.

22. The envelope surrounding the coat of a virus comes from the <u>host</u>.

23. Hepatitis is a <u>viroid</u> disease.

24. Viroids are <u>smaller</u> than viruses.

25. Viroids are known to cause several <u>animal</u> diseases.

Answers to Questions for Review

PART A

1. membrane-bound
2. cholesterol
3. mesosome
4. cell wall
5. murein
6. ribosome
7. replication
8. bacteria
9. species
10. eubacteria
11. binary fission
12. saprobe
13. bacilli
14. nitrifying
15. aerobes
16. facultative anaerobes
17. purple
18. metabolic
19. endospores
20. prochlorophyta
21. heterocysts
22. fresh
23. virion
24. bacteriophage
25. phytoplankton

PART B

1. c	10. a	19. a
2. a	11. d	20. b
3. d	12. a	21. c
4. c	13. d	22. a
5. d	14. c	23. d
6. c	15. b	24. c
7. b	16. b	25. b
8. d	17. c	
9. c	18. d	

PART C

1. true
2. invisible
3. mycoplasmas
4. ribosome
5. $\frac{1}{1000}$
6. one half
7. Chemosynthetic
8. 20 minutes
9. capsule
10. nodules
11. cocci
12. intestine
13. fermentation
14. oxygen
15. true
16. true
17. helpful
18. bacteria
19. rope
20. carbohydrate
21. anaerobic
22. true
23. virus
24. true
25. plant

CHAPTER 5

THE PROTIST KINGDOM: PROTOZOA AND UNICELLULAR ALGAE

All of the species assigned to the kingdom Protista are eukaryotic. Most protists carry out their lives within a single cell as free-living organisms. However, some protist species are organized into *colonies* where each cell carries out its own life functions and where, also, there may be some simple division of labor among the cells in the grouping. An impressive variety of species are classified as protists, and they probably descended from diverse evolutionary lines. The protists themselves represent evolutionary modification and are probably the ancestors of the modern fungi, plants and animals.

Major Groups of Protists

Most modern classification schemes divide the Protista into three major groups: the protozoa, or animal-like protists; the fungus-like protists; and the plant-like protists.

PROTOZOA, THE ANIMAL-LIKE PROTISTS

Protozoa, meaning "first animals," are one-celled heterotrophs. Species of protozoa number in the thousands. They live in fresh water, salt water, dry sand and moist soil. Some species live as parasites on or inside of the bodies of other organisms. Reproduction in the protozoans is usually described as being asexual by means of mitosis, but recent research has revealed that many protozoa augment asexual reproduction with a sexual cycle. Usually, the sexual cycle occurs during periods of adverse environmental conditions, and the cell arising from the fusion of gametes (*zygote*) can resist unfavorable conditions. The thick wall and the decreased metabolic rate of the *cyst* permits survival during periods of cold, drought or famine.

The protozoa are divided into four phyla, based primarily on the methods of locomotion.

69

Mastigophora

The Mastigophora are protozoa that have one or more flagella. This phylum, also known as the Zoomastigina or zooflagellates, includes a rather heterogeneous group of organisms. Some species are free-living and inhabit fresh or salt water; of these, some are free-swimming; others glide over the surface of rocks; and some are sessile, attached to available submerged surfaces. Other Mastigophora species live in a *symbiotic* relationship with organisms of other species, each helping the other with a particular life function. For example, several species live in the intestines of termites, cockroaches and woodroaches, where they digest cellulose for these insects. The genus *Trypanosoma* includes parasites that cause debilitating diseases in human beings. *Trypanosoma gambiense* (Fig. 5.1) is the zooflagellate that causes African sleeping sickness. Humans are infected with the trypanosome by the bite of an infected tsetse fly.

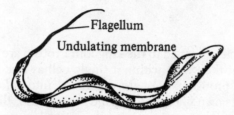

Flagellum
Undulating membrane

Fig. 5.1 The causative organism of African sleeping sickness, *Trypanosoma gambiense*

Sarcodina

The members of the phylum Sarcodina are described as being amoeboid. *Amoeba proteus* (Fig. 5.2) is the type species. Species included in the Sarcodina move by means of *pseudopods*, flowing extensions of the flexible and amorphous body. The pseudopods also serve in food-catching. Most of the sarcodines live in fresh water. A *contractile vacuole*, an organelle designed to expel excess water from the protist cell body, plays an important role in maintaining water balance. Food is temporarily stored in a food vacuole where it is digested by the action of enzymes.

The Foraminifera and the Radiolaria are groups of marine sarcodines that secrete hard shells of mineral compounds around themselves. When they die, their shells become an important constituent of the bottom mud of the ocean floor. The Foraminifera have built the limestone and chalk deposits that date back to the Cambrian period; the Radiolarians, the siliceous rocks dating back to the Precambrian period.

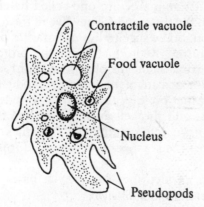

Contractile vacuole

Food vacuole

Nucleus

Pseudopods

Fig. 5.2 *Amoeba proteus*

Sporozoa

The Sporozoa are parasitic spore-formers. The adult forms are incapable of locomotion, although immature organisms may move by means of pseudopodia. Some species of sporozoa go through a complicated life cycle requiring different hosts during different life stages. For example, the species *Plasmodium vivax*—the agent that causes malaria—requires two hosts: the *Anopheles* mosquito and a human (Fig. 5.3).

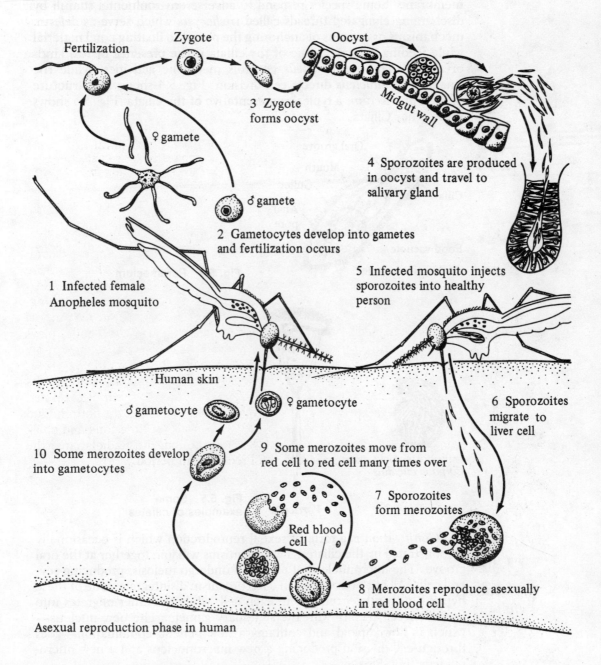

Sexual reproduction phase in mosquito

Fertilization

Zygote

Oocyst

♀ gamete

Midgut wall

3 Zygote forms oocyst

♂ gamete

4 Sporozoites are produced in oocyst and travel to salivary gland

2 Gametocytes develop into gametes and fertilization occurs

1 Infected female Anopheles mosquito

5 Infected mosquito injects sporozoites into healthy person

Human skin

♂ gametocyte

♀ gametocyte

6 Sporozoites migrate to liver cell

10 Some merozoites develop into gametocytes

9 Some merozoites move from red cell to red cell many times over

7 Sporozoites form merozoites

Red blood cell

8 Merozoites reproduce asexually in red blood cell

Asexual reproduction phase in human

Fig. 5.3 Life cycle of *Plasmodium vivax*

Ciliata

Of the four protozoan divisions, the phylum Ciliata has the greatest number of species. Species belonging to this phylum have *cilia*, short cytoplasmic strands which are used for locomotion and in some cases to sweep food particles into an opening called the *oral groove*. Protozoans included in this phylum inhabit fresh water and salt water. Some are free-swimming; some creep; others are sessile; and some are parasitic in other animals.

The cytoplasm in ciliates is differentiated into rigid outer ectoplasm and a more fluid inner endoplasm. A *pellicle* lies just inside of the cell membrane. Some species respond to adverse environmental stimuli by discharging elongated threads called *trichocysts* which serve as defense mechanisms or a means of anchoring the protist to floating pond material while feeding. Characteristic of the ciliata is the presence of two kinds of nuclei. The *macronucleus* controls metabolic activities, while the smaller *micronucleus* directs cell division. Fig. 5.4 shows the structure of the *Paramecium*, a typical representative of the ciliata. Fig. 5.5 shows some other Ciliata.

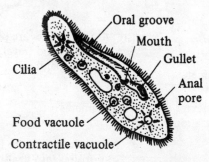

Fig. 5.4 Paramecium

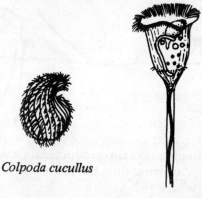

Colpoda cucullus

Vorticella

Fig. 5.5 Some examples of ciliates

Conjugation is a form of sexual reproduction which is occasionally demonstrated by the ciliates. Two organisms will join together at the oral groove. The micronucleus of each will undergo meiosis, producing several cells. All but two of these in each organism disintegrate. One of these haploid micronuclei remains in each cell, while the other migrates into the other cell, fusing with the stationary gamete. The new nucleus—which is now diploid and contains a new genetic combination—goes through cell division producing a new macronucleus and a new micronucleus (Fig. 5.6).

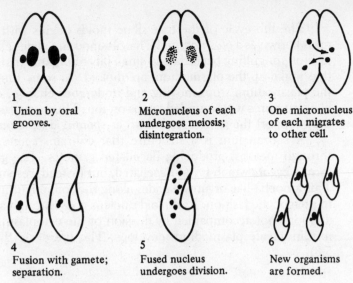

1
Union by oral
grooves.

2
Micronucleus of each
undergoes meiosis;
disintegration.

3
One micronucleus
of each migrates
to other cell.

4
Fusion with gamete;
separation.

5
Fused nucleus
undergoes division.

6
New organisms
are formed.

Fig. 5.6 Conjugation in paramecia

THE FUNGUS-LIKE PROTISTS

A *fungus* is an organism that obtains food by *absorbing* it from dead organic matter or from the body of a living host. There are two groups of organisms that are considered protists and yet their way of life and mode of nutrition are like those of the true fungi. These funguslike protists are the *Protomycota* and the *Gymnomycota*.

Protomycota

The Protomycota are small, unicellular organisms not readily visible. Many of them live as parasites or saprophytes on water-dwelling plants. A few species live in the soil. Others live as parasites inside of invertebrates such as liver fluke, nematodes or mosquito larvae.

The chytrids and the hypochytrids—two representative groups—live in association with a host organism or cell. One form of chytrid is a haploid cell that lives within another cell. At a particular time the nucleus goes through a series of mitotic divisions without division of the cytoplasm. When the nuclear divisions are completed, a little of the cytoplasm surrounds each nucleus. Each new cell develops a flagellum and these motile *zoospores* are released into the surrounding water (Fig. 5.7). Some may fuse and go through a sexual reproduction. Others merely develop into another vegetative cell and then the process repeats. Chytrids show various adaptations for absorbing nutrients. Some have cytoplasmic extensions known as *rhizoids* specialized for absorption.

Gymnomycota—Slime Molds

Vegetative thallus   Zoospore

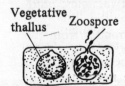

Fig. 5.7 Vegetative chytrid cells

There are two major groups of slime molds: the Myxomycota, or the true *slime molds*, and the Acrasiomycota, the *cellular slime molds*.

True Slime Molds

Slime molds live on the forest floor where they grow in damp soil, on or around rotting logs and on decaying vegetation. They appear to be shapeless globs of slime of varying colors: white, yellow or red.

The life cycle of the true slime molds begins with a multinucleate mass known as a *plasmodium*. This plasmodium glides about in amoeboid fashion, engulfing bacteria and small bits of organic material. The nuclei that make up the plasmodium are diploid. At some time in its life cycle the plasmodium stops moving and undergoes change, developing stalk-like structures with rounded knobs on top. These are *fruiting bodies*, and they support the structures known as *sporangia* (*sporangium*, sing.).

A sporangium is a structure that contains *spores*. The spores go through meiosis, producing flagellated gametes. The gametes fuse and form a *zygote* which is not flagellated, but instead, resembles an amoeba. This amoeba-like organism glides along the soil, engulfing food materials in phagocytic fashion. Its diploid nucleus goes through a series of mitotic divisions not accompanied by division of the cytoplasm. In this way the mutlinucleate plasmodium develops. The life cycle then repeats (Fig. 5.8).

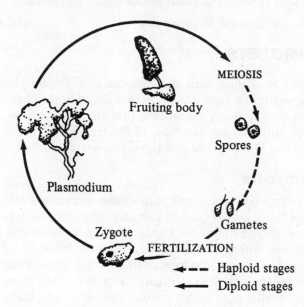

Fig. 5.8 Life cycle of a true slime mold

Cellular Slime Molds

The life cycle of the cellular slime molds, often called the social amoebae, is much different from that of the true slime molds. The fruiting bodies of the social amoeba are called *sorocarps*. The spores are carried by wind, water, insects or other means to new environments. Those spores placed in favorable environments will germinate. As the spore wall disintegrates, an amoeba, using pseudopods, pushes its way out onto the soil. These amoebae resemble other amoebae in structure. Each has a cell membrane, a nucleus, nucleolus, contractile vacuoles, food vacuoles, mitochondria and endoplasmic reticulum.

The free-living haploid amoebae feed on organic matter from the soil and grow to an optimum size. They then divide by mitosis and cytokinesis, producing daughter cells which, like the parent, have one haploid nucleus. Then their behavior changes. They stop feeding and begin to move in a directed manner toward definite centers or collecting points where they form closely packed groups and take on the formation of a compact mass of cells.

Next they become grouped together forming a *pseudo-plasmodium* or slug. A very thin sheath of polysaccharide material surrounds the mass, but the cells maintain their individuality. The finger-shaped pseudo-plasmodium moves across the soil substrate slowly but in a directed and coordinated way. Eventually, the front end of the pseudoplasmodium stops moving and the rear segment moves underneath the front end to form a mound of cells. At this time, differentiation of cells begins to take place. Stalk cells which give rise to a fruiting body are formed, and the life cycle repeats (Fig. 5.9).

PLANT-LIKE PROTISTA

There are three major groups of plant-like protists.

The Euglenophyta

The euglenoids are represented by the organism *Euglena* (Fig. 5.10). This unicellular organism has both plant-like and animal-like characteristics. It has chlorophyll *a* and *b* and some carotenoids, and it is able to carry on photosynthesis. However, *Euglena* lacks a cell wall, swims by using a flagellum, has a light-sensitive red-orange eyespot known as the *stigma* and a large contractile vacuole. *Euglena* also has a *pyrenoid body* that functions in the synthesis of *paramylum*, a carbohydrate storage product that is peculiar to the euglenoids. These organisms reproduce asexually by longitudinal mitotic cell division. However, during mitosis, the cell membrane remains intact and the nucleolus persists.

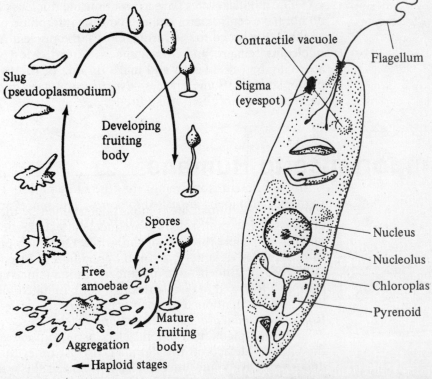

Slug (pseudoplasmodium)

Developing fruiting body

Spores

Free amoebae

Aggregation

Mature fruiting body

← Haploid stages

Fig. 5.9 Life cycle of a cellular slime mold

Contractile vacuole

Flagellum

Stigma (eyespot)

Nucleus

Nucleolus

Chloroplas

Pyrenoid

Fig. 5.10 Euglena

Fig. 5.11 A chrysophyte

The Chrysophyta

The yellow-green algae, the golden-brown algae and the diatoms are included in this group. Most species in the Chrysophyta group are single-celled and reproduce asexually. There are a few simple multicellular forms. The members of this group share certain characteristics. They are pigmented with chlorophyll *a* and *c* (lacking *b*) and contain the carotenoid *fucoxanthin*, which gives them the golden color. These protists live in salt water, fresh water, and in damp places between rocks. They use a polysaccharide called chrysolaminaran instead of starch. Many of the Chrysophyta have two flagella of unequal length at the anterior end of the body (Fig. 5.11); but some have only one flagellum and others, none.

The diatoms are diploid cells that lack flagella. Their cell walls, composed of two pieces, one fitting over the other, are impregnated with silica and pectin and often have intricate patterns in the forms of pits and ridges. When the organisms die, the shells fall to the bottom of the water where they disintegrate and form *diatomaceous earth*, a substance used as an abrasive in silver polish and detergents, as the packing in air and water filters, and in paint removers, deodorizing oils and fertilizers.

The Pyrrophyta

The dinoflagellates are small protists and usually unicellular. Most of these organisms have two unequal flagella, one extending longitudinally from the posterior end of the cell, the other encircling the central part of the cell. Some dinoflagellates extend trichocysts like the *Paramecium*; others have *nematocysts, stinging cells* common in the coelenterates.

The dinoflagellates have a most unusual nucleus. The chromosomes do not have centromeres and even during interphase remain in evidence as short, thickened rods. During mitosis, the nuclear membrane and the nucleolus remain, and no spindle is formed. Most species reproduce asexually by cell division and make up one of the two main groups of phytoplankton. Some species—*Noctiluca*, for example—are bioluminescent, giving off light like a firefly.

Importance to Humans _____

The quality of human life is in part determined by the ability to be free of disease. The protist disease-producers affect the health of large populations of people throughout the world. One such protist-initiated disease is amebic dysentery, a painful condition of bleeding ulcers caused by a type of amoeba that parasitizes human intestines. This protist-pathogen is passed to people through contaminated food and water. We have already mentioned two pathogenic protists carried by insects—*Trypanosoma gambiense* and *Plasmodium vivax*.

The red pigmented dinoflagellate *Gonyaulax* live in the waters of the Gulf of Mexico off the Florida coast. At times this organism reproduces explosively and uncontrollably coloring the waters red. This so-called "red tide" poisons millions of fish and does harm to people who eat these fish.

Many other protist species are helpful to humans. Slime molds help to keep an ecological balance by feeding on decayed plant and animal matter. These organisms also have a great deal of value as research specimens for investigators who are trying to ferret out the secrets of cell specialization and differentiation.

Plankton is the mass of green that floats on rivers, lakes and oceans. In reality, protists—microscopic "plant" life (phytoplankton) and microscopic "animal" life (zooplankton)—intermingle with mutual benefit in this floating mass which serves vital roles in the food chains of aquatic species.

Chronology of Famous Names in Biology

1667   **Anton Leeuwenhoek** (Dutch)—was the first person to see and describe protozoa.

1786   **Otto Frederick Muller** (Danish)—wrote the first treatise on protozoa titled *Animalcula Infusoria*.

1836   **Christian G. Ehrenberg** (German)—wrote a beautifully illustrated book describing 69 protozoa accurately.

1841   **Felix Dujardin** (French)—discovered protoplasm in protozoa.

1845   **Friedrich Stein** (Austrian)—named the orders and suborders of protozoa.

1880   **Richard Hertwig** (German)—discovered chromatin in the protozoan nucleus.

1959   **John T. Bonner** (American)—made a number of investigative studies which elucidated basic facts about the behavior and structure of cellular slime mold.

1962   **John T. Bonner** (American)—discovered that cAMP directs the movement of the social amoeba toward a collection center.

Words for Study

chytrid	contractile vacuole	dinoflagellate
cilia	cyst	hypochytrid
colony	diatom	macronucleus
conjugation	diatomaceous earth	micronucleus

nematocyst	pseudoplasmodium	stigma
oral groove	pseudopod	symbiosis
paramylum	pyrenoid body	trichocysts
pellicle	rhizoid	tsetse
plankton	slime mold	zooflagellate
plasmodium	social amoeba	zooplankton
protist	sorocarp	zoospore
protozoa	sporangium	zygote

Questions for Review

PART A.   Completion. Write in the word that correctly completes the statement.

1. One-celled protists that resemble animal cells are the ..1..
2. Another name for a "self-feeder" is a(an) ..2..
3. *Amoeba* move by means of false feet known as ..3..
4. The usual mode of reproduction in protozoa is ..4..
5. The relationship in which two organisms of different species live together and neither is harmed by the association is known as ..5..
6. Digestion in *Amoeba proteus* takes place in the ..6..
7. The Sporozoa are harmful to organisms of other species and are therefore classified as ..7..
8. *Anopheles* is the genus name of a ..8..
9. The organelle that expels excess water from the protist is the ..9..
10. In the *Paramecium*, metabolic activity is controlled by which nucleus?
11. The body of the paramecium is prevented from being totally flexible by the ..11..
12. The form of reproduction in which like gametes fuse is called ..12..
13. The plasmodium is a stage in the life cycle of ..13..
14. A spore-producing structure is known as a ..14..
15. Sorocarps are correctly associated with the ..15.. slime molds.
16. In *Euglena*, the stigma is sensitive to ..16..
17. The cell walls of the diatoms are impregnated with pectin and ..17..
18. Diatomaceous earth forms from the ..18.. of diatoms.
19. The number of flagella usually found in dinoflagellates is ..19..
20. The number of flagella usually present in diatoms is ..20..

PART B.   Multiple Choice. Circle the letter of the item that correctly completes each statement.

1. Classification systems are best described as being
   (a) unchanging   (c) unreliable
   (b) fixed   (d) artificial

2. The evolution of protists involved the development of
   (a) membrane bound organelles (c) functioning mitochondria
   (b) a discrete nucleus   (d) an active Golgi

3. Protists live
   (a) on land   (c) in fresh water
   (b) in the sea   (d) in all of these

4. The primary purpose of the protozoan cyst is to
   (a) produce new organisms
   (b) survive unfavorable conditions
   (c) increase the metabolic rate
   (d) aid in better nutrition

5. *Trypanosoma gambiense* is the organism that causes
   (a) botulism
   (b) cholera
   (c) African sleeping sickness
   (d) malaria

6. Limestone and chalk deposits are found in the shells of dead
   (a) diatoms   (c) Foraminifera
   (b) Sarcodina   (d) zooflagellates

7. *Plasmodium vivax* is the organism that causes
   (a) botulism   (c) African sleeping sickness
   (b) dysentery   (d) malaria

8. Trichocysts in ciliates serve mainly
   (a) as defense mechanisms (c) to expel water
   (b) to gather food   (d) for movement

9. Structures used for locomotion in the protists include all of the following except
   (a) pseudopodia   (c) cilia
   (b) legs   (d) flagella

10. The best way to prevent malaria is by destroying the breeding places of
   (a) *Plasmodium vivax*   (c) tsetse flies
   (b) *Trypanosoma vivax* (d) *Anopheles mosquitos*

11. *Paramecia* move by means of
   (a) flagella   (c) trichocysts
   (b) cilia   (d) pseudopodia

12. Flagellated gametes produced by chytrids are the
   (a) zygotes   (c) zoospores
   (b) zooflagellates   (d) zymogens

13. Rhizoids are structures specialized for
   (a) reproduction   (c) locomotion
   (b) absorbing nutrients (d) providing rigidity

14. The plasmodium of true slime molds is best described as
   (a) multinucleate   (c) binucleate
   (b) uninucleate   (d) prokaryotic

15. In part of the life cycle of the true slime molds, sporangia are sup-
    ported by structures known as
   (a) hyphae   (c) rhizoids
   (b) mycelia   (d) fruiting bodies

16. During mitosis in *Euglena*
   (a) the nucleolus disappears
   (b) chromosomes do not form
   (c) the nuclear membrane persists
   (d) transverse fission occurs

17. The yellow-green algae lack
   (a) chlorophyll *a*   (c) chlorophyll *c*
   (b) chlorophyll *b*   (d) caratenoids

18. An organism that lives inside of another and does harm to its host is
    known as a (an)
   (a) autotroph   (c) saprophyte
   (b) symbiont   (d) parasite

19. *Noctiluca* is noted for its
   (a) reproduction mode   (c) swimming ability
   (b) bioluminescense   (d) apparent immortality

20. Nematocysts are used for
   (a) swimming   (c) gliding
   (b) slinging   (d) stinging

PART C.   **Modified True-False.** If a statement is correct, write "true"
          for your answer. If a statement is incorrect, change the
          underlined word to one that will make it correct.

1. Protists are prokaryotes.

2. Chromosomes in the eukaryotes are best described as circular.

3. Protists descended from the same evolutionary lines.

4. Protozoa are unicellular autotrophs.

5. Protozoa are classified according to methods of nutrition.

6. *Trypanosoma gambiense* is carried by the insect known as the house
   fly.

7. *Trypanosoma gambiense* is best classified as a dinoflagellate.

8. *Plasmodium vivax* requires two hosts: fly and humans.

9. Adult sporozoans are motile.

10. The pseudoplasmodium of social amoebae moves in a directed way.

11. In *Paramecia*, reproduction is controlled by the macronucleus.

12. Chytrids are best classified as fungi.

13. The nuclei in the plasmodium of the true slime molds is haploid.

14. The pyrenoid body in *Euglena* stores <u>glycogen</u>.

15. The chloroplasts enable *Euglena* to carry out <u>heterotrophic</u> nutrition.

16. The pigment fucoxanthin causes the <u>green</u> color in the chrysophyta.

17. *Euglena* swims by means of a <u>flagellum</u>.

18. The diatoms have <u>no</u> commercial value.

19. Water balance in the amoeba is controlled by the <u>food</u> vacuole.

20. The chromosomes in the dinoflagellate nucleus lack structures called <u>chromatids</u>.

Answers to Questions for Review

PART A

1. protozoa
2. autotroph
3. pseudopodia
4. asexual
5. symbiosis
6. food vacuole
7. parasites
8. mosquito
9. contractile vacuole
10. macronucleus
11. pellicle
12. conjugation
13. slime molds
14. sporangium
15. cellular
16. light
17. silica
18. cell walls
19. two
20. zero

PART B

1. d
2. a
3. d
4. b
5. c
6. c
7. d
8. a
9. b
10. d
11. b
12. c
13. b
14. a
15. d
16. c
17. b
18. d
19. b
20. d

PART C

1. eukaryotes
2. linear
3. diverse
4. heterotrophs
5. locomotion
6. tsetse
7. zooflagellate
8. mosquito
9. nonmotile
10. true
11. micronucleus
12. protists
13. diploid
14. paramylum
15. autotrophic
16. yellow
17. truc
18. great
19. contractile
20. centromeres

CHAPTER 6

THE FUNGI

Most fungi are eukaryotic, multicellular and multinucleate organisms. Yeasts are unicellular forms. The cells of fungi are different from those of other species because the boundaries separating the cells are either entirely missing or only partially formed. Thus fungi are primarily *coenocytic* organisms; this means that the cells have more than one nucleus in a single mass of cytoplasm. However, the characteristic that most distinguishes the fungi from other organisms is their mode of nutrition.

General Features

NUTRITION

Since fungi do not have chlorophyll, they cannot produce their food autotrophically by photosynthesis. Fungi are heterotrophs. However, they cannot engulf food phagocytically as do the amoeba, nor can they ingest food as do animals equipped with a mouth. Some fungi are restricted to a saprophytic way of life, obtaining nutrition by absorbing dead organic matter, while others are parasitic, absorbing their nutrition from living hosts. In either case, fungi must live very near their food supply in order to stay alive. Specialized fungal structures secrete digestive (hydrolytic) enzymes into the food substrate. The organic molecules of the substrate are made smaller by these enzymes and thus can be absorbed by the fungus.

BODY ORGANIZATION

Fig. 6.1 Spore

The fungus begins its life as a *spore* (Fig. 6.1). A spore is a microscopic cell that, under favorable conditions, will develop into a new individual. The cell wall of the spore is thick and tough, resistant to adverse environmental conditions. However, the spore is light in weight and is trans-

Fig. 6.2 Germinating spore

Fig. 6.3 Mycelium

Fig. 6.4 Internal view of a hypha

ported to new habitats by currents of air. When a spore settles on its substrate in favorable environmental conditions, its cell wall breaks open and the spore commences to sprout or *germinate* (Fig. 6.2). The germinating spore absorbs food and grows an elongated thread called a *hypha*. As growth continues, many hyphae develop until they appear as a tangled mass of threads. When many hyphae appear, the body of the fungus is now called a *mycelium* (Fig. 6.3). A hypha gives off a chemical that makes other hyphae grow away from it. In this way, competition for food is reduced among hyphae. In general, the function of the mycelium is to absorb food and to produce new fungus plants.

A microscopic study of a hypha provides interesting information about its internal structure. Some hyphae are multinucleate with many nuclei sharing the same mass of cytoplasm without dividing membranes (Fig. 6.4). Other hypha are divided into compartments by *septa*. These compartments contain one or two nuclei.

Parasitic fungi have special structures called *haustoria* that are specialized to penetrate the cell walls of plants. The haustoria grow into plant cell cytoplasm and absorb nutrients directly from the cells which they parasitize (Fig. 6.5).

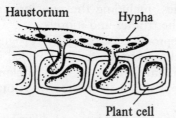

Fig. 6.5 Haustoria

METHODS OF REPRODUCTION

Reproduction in fungi usually occurs by the production of spores. Special reproductive hyphae called *sporangiophores* have growing at their tips *sporangia*, (sing. sporangium) or spore cases. It is within the spore case that reproductive spores form.

Some fungi also reproduce vegetatively. V*egetative reproduction* is the process by which a new individual is produced from a part of the parent's body without involving sex cells. New individuals produced vegetatively are identical to the parent. Broken fragments of mycelia produce new fungus individuals identical to themselves.

In most fungi asexual reproduction (vegetative and sporulation) is augmented by a sexual cycle. The sexual phase begins with the fusion of *gametangia* (gametangium, sing.), special gamete-producing structures. This is followed by a fusion of special nuclei or gametes resulting in the formation of a zygote.

Major Groups of Fungi

The phylum Mycota is the only phylum in the kingdom Fungi. It is divided into five classes: Oomycetes, Ascomycetes, Zygomycetes, Basidiomycetes, and Deuteromycetes.

CLASS OOMYCETES

The Oomycetes are primarily water molds, although some species live on land. This is the only fungus class in which the cell walls are made of cellulose, not of chitin, and in which the gametes are differentiated into male sperm and female egg cells. Another characteristic of the class is that the spores are flagellated and require free water for swimming.

Most of the Oomycetes are *saprobes*, absorbing their nutrients from dead organic matter. Some species are parasitic and disease-producing. For example: *Albugo candida* causes white rust on cabbage and other leafy plants. *Saprolegnia*, a saprobe, grows as mold on the water-borne bodies of dead insects, fish and frogs.

CLASS ASCOMYCETES

Yeast is an example of a single-celled member of the Ascomycetes. Yeast cells are small, oval-shaped structures that reproduce by budding. Most Ascomycetes are multicellular, however. This is the largest class of fungi with about 30,000 species, including the powdery mildews, black and blue-green molds, and the truffles and morels.

The Ascomycetes reproduce asexually by very fine spores known as *conidia*. The hypha of the Ascomycetes are divided into compartments by septa. Each compartment contains its own nucleus, but pores in the septa allow the migration of cell structures from one compartment to another.

The Ascomycetes are so-named because during part of the life cycle reproductive cells are held in a little sac or *ascus*. At a certain time, two hyphae grow together. Although their cytoplasm intermingles, the nuclei remain separate and do not fuse. The new hyphae that grow from this fused structure have nuclei of different genetic strains. This is known as a *dikaryon*. The dikaryon fuses with nonreproductive hyphae to form a fruiting body in which the asci form. At this time, within the cell that is to become the ascus, the two nuclei of the dikaryon fuse. They now go through a series of meiotic and mitotic divisions, resulting in eight haploid nuclei, which soon become surrounded by their own walls to form eight *ascospores* that disperse when the ascus ruptures.

CLASS ZYGOMYCETES

Most of the Zygomycetes are land-dwelling organisms that inhabit the soil and carry out a saprobic way of life. Their hyphae are coenocytic and the cell walls are composed of *chitin*. Chitin is a tough, nitrogen-containing polysaccharide that is present in the exoskeletons of insects and in the cell walls of most species of fungi.

The type species of this class is *Rhizopus stolonifer* (Fig. 6.6), the black bread mold. The life cycle of *Rhizopus* begins with a germinating spore that is established on a favorable substrate such as a piece of bread in a moist, warm, dark environment. Hyphae grow from the spore, eventually forming a mycelium. Some of the hyphae become specialized into rhizoids, downward growing threads that secure the mycelium to the substrate. The rhizoids have additional functions, also. They send out

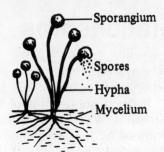

Fig. 6.6 Bread mold and spores

digestive juices which break down the large organic molecules of the substrate into smaller ones in the process of digestion. The rhizoids then perform their third function by absorbing the digested material. Other specialized hyphae grow upward. These are known as *sporangiophores* because they bear at their tips sporangia which produce haploid spores.

During part of its life cycle, *Rhizopus* goes through sexual reproduction. Hyphae of opposite mating strains form a *gametangium*, a structure that produces special nuclei that function as gametes. The gametangia of opposite mating strains fuse. Some of their nuclei pair off and fuse forming diploid nuclei. A thick wall covers the fused gametangia and their zygote nuclei to form a *zygospore*. The zygospore remains dormant for a while. When the zygospore germinates, meiosis takes place in the zygotes. An aerial hypha grows from the activated zygospore. This hypha is a sporangiophore and supports a sporangium which produces many haploid spores. The life cycle repeats.

CLASS BASIDIOMYCETES

The Basidiomycetes include the mushrooms, bracket fungi and the smuts. The life cycle of the Basidiomycetes begins when a haploid spore germinates, giving rise to hyphae. Hyphae of different mating types fuse forming mycelia that are dikaryotic. Most of the life history of the mushrooms and its relatives is spent in the dikaryotic stage.

The name of this class is derived from the *basidium* (club) that forms at the tip of each reproductive hypha. Inside each basidium, the two haploid nuclei fuse, forming a diploid nucleus. This nucleus undergoes meiosis, thereby producing four haploid nuclei. The haploid nuclei move to the outer edge of the basidium. Each nucleus becomes surrounded by an elongated cell wall in a structure known as the *basidiospore*. The basidiospore is maintained on a delicate stalk which separates it from the rest of the fungus. Dissemination of the spore takes place when the stalk breaks and is carried away by the wind.

CLASS DEUTEROMYCETES

This class contains those species known as the "Fungi Imperfecti". Those species that puzzle taxonomists are put into this group. Most of these organisms have a life cycle that is not typical of fungi. Since sexual reproduction has not been observed in these species, they are called imperfect fungi. The organisms that cause the human diseases of ringworm, athlete's foot and thrush are classed as deuteromycetes.

Special Nutritional Relationships _____

Fungi are often found associated with other species in special relationships. When two different species of organisms live together, the relationship is called *symbiosis*. If the relationship is of mutual benefit to both species, it is called *mutualism*. When one species benefits and the other does not but is not harmed by the association, the condition is known as *commensalism*. When one species lives at the expense of another, doing harm to its host, the relationship is *parasitism*. Disease-producing organisms are parasites.

LICHENS

Lichens are pioneer organisms that can inhabit bare rock and other uninviting substrates. They live on the barks of trees and even on stone walls. A lichen is a combination of two organisms—an alga and a fungus—that live together in a mutualistic relationship. The alga carries on photosynthesis, while the fungus absorbs water and mineral matter for its partner. The fungus also anchors the lichen to the substrate. Scientists have determined that the alga in this partnership can live alone. The fungus, which may be an ascomycete, imperfect fungus, or basidiomycete, cannot exist by itself and is therefore the dependent member of the team. New lichens are formed by the capture of an alga by a fungus. If the fungus kills the alga, the fungus also dies.

MYCORRHIZA

Several fungus species, including mushrooms, live in close association with plant roots. The mushroom absorbs minerals from the soil and passes them along to the plant on which it lives. This condition is known as *mycorrhiza*, meaning "fungus root". Scientists are not too sure how the plant helps the fungus, but it is known that most plant families grow better when in association with fungi.

Importance to Humans _____

From the viewpoint of environment and ecology, fungi help to keep the natural environment in balance. Species dependent on dead organic matter as their source of nutrition assist with the breaking down of fallen leaves, the dead bodies of plants and animals, and animal wastes. Fungi have the ability to absorb moisture from the air, a characteristic that permits them to live in environments that do not have adequate soil water as required by other species.

Most fungi are not pathogenic to humans. However, some nonparasitic fungi do work against human interests. There are fungus species that live quite well digesting such unlikely substrates as the insulation on telephone wires, leather, polyvinyl plastics, cork, hair and wax. There are those species that cause the mildew of clothing, wall paper and books. Then there are those wood-rotting fungi that destroy the wood construction in houses and ships.

Parasitic fungi invade plants more readily than animal bodies. Diseases of food crops can have disastrous effects on human populations. This occurred when the oomycete *Phytophthora infestans* caused the potato blight in Ireland, resulting in a devastating famine between the years 1845 and 1851 and when *Plasmopara viticola,* the cause of downy mildew of grapes, nearly ruined the wine-making industry of France. Ergot, a disease of rye, is caused by the ascomycete *Claviceps purpurea.* Ergotism, the disease that affects humans who eat the infected rye, induces gangrene, nervous spasms, convulsions, and psychotic delusions. Ergot is the source of the hallucinogen LSD. Other plant diseases caused by the Ascomycetes are peach leaf curl, Dutch elm disease, chestnut blight and apple scab.

Some Ascomycetes are beneficial to humans. These include the truffles and the morels, edible fruiting bodies, that cost $400.00 a pound in New York. *Saccharomyces cerevisiae* is the species of yeast needed in the fermenting processes of malt, barley and hops to make beer. The fermenting of grapes produces wine. Another species of yeast is used in making dough.

The Fungi Imperfecti include *Penicillium notatum*, the source of the antibiotic penicillin. This group of fungi also includes those species that help to ripen Roquefort and Camenbert cheeses. Soy sauce is prepared by fermenting soybeans with the mold *Aspergillus oryzae.*

Chronology of Famous Names in Biology

1929 **Alexander Fleming** (English)—discovered the antibiotic properties of the fungus *Penicillium notatum.*

1943 **Albert Hoffman** (Swiss)—isolated LSD from ergot and discovered the powerful hallucinogenic properties.

Words for Study

ascus	gametangium	saphrophyte
ascospore	haustoria	saprobe
basidiospore	host	septa
basidium	hypha	sporangia
chitin	lichen	sporangiophores
coenocytic	mutualism	spore
commensalism	mycelium	symbiosis
conidia	mycorrhiza	vegetative reproduction
dikaryon	parasite	zygospore
germinate	rhizoid	

Questions for Review

PART A.   Completion. Write in the word that correctly completes each statement.

1. A cell that has several nuclei in a single mass of protoplasm is called a ..1..
2. Fungi absorb dissolved food molecules from dead organic matter and thus are classified as ..2..
3. Hydrolytic enzymes are the same as ..3.. enyzymes.
4. Vegetative reproduction does not involve ..4.. cells.
5. Conidia are cells known more commonly as ..5..
6. An example of a unicellular ascomycete is ..6..
7. Rhizoids secrete substances that serve the purpose of ..7..
8. The combination of an alga and a fungus living in a mutualistic relationship is called a ..8..
9. Ringworm is a ..9.. infection and not one caused by worms.
10. A cell that has two nuclei originating from different genetic strains is known as a ..10..

PART B.   Multiple Choice. Circle the letter of the item that correctly completes each statement.

1. A saprobe is an organism that
   (a) absorbs material from living cells
   (b) absorbs material from dead cells
   (c) ingests dead organic matter
   (d) ingests living organic matter

2. Septa are
   (a) compartments
   (b) double nuclei
   (c) dividing walls
   (d) masses of protoplasm

3. Gametangia are assoiated with
   (a) sporulation
   (b) vegetative reproduction
   (c) germination
   (d) sexual reproduction

4. The only fungus class in which the gametes are differentiated into sperm and egg is the
   (a) Oomycetes
   (b) Zygomycetes
   (c) Ascomycetes
   (d) Basidiomycetes

5. A dikaryon is a cell that has
   (a) a single nucleus and lots of cytoplasm
   (b) perforated cell membranes
   (c) two nuclei of different genetic strains
   (d) a micronucleus and a macronucleus

6. Zygospores are characteristic of the fungus class
   (a) oomycetes
   (b) Fungi Imperfecti
   (c) Basidiomycetes
   (d) Zygomycetes

7. Black bread mold is the type species for the fungus class
   (a) Oomycetes   (c) Deuteromycetes
   (b) Basidiomycetes   (d) Zygomycetes

8. Each class in the kingdom Fungi uses special methods to produce
   (a) hypha   (c) mycelia
   (b) rhizoids   (d) spores

9. Fungi Imperfecti are so named because
   (a) sexual reproduction has not been observed in them
   (b) they cause human disease
   (c) many of their hyphae are missing
   (d) they do not reproduce by spores

10. The disease of rye which is devastating to humans is known as
    (a) morel   (c) ergot
    (b) truffle   (d) ringworm

PART C.   Modified True-False. If a statement is true, write "true" for your answer. If a statement is incorrect, change the underlined word to one that will make the statement true.

1. Most of the fungi begin life as a zygote.

2. To germinate means to contaminate.

3. Flagellated swimming spores are characteristic of the fungus class Zygomycetes.

4. Ascospores are diploid.

5. A nitrogen-containing polysaccharide that composes the ccll walls of most fungi is cellulose.

6. *Rhizopus stolonifer* is the scientific name for blue-green mold.

7. When two gametes fuse, a zygospore is formed.

8. An example of the class Basidiomycetes is watermold.

9. The word basidium means mold.

10. Pioneer organisms that can inhabit bare rocks are lichens.

Answers to Questions for Review

PART A

1. coenocyte
2. saprobes
3. digestive
4. sex
5. spores

6. yeast
7. digestion
8. lichen
9. fungus
10. dikaryon

PART B

1. b	5. c	8. d
2. c	6. d	9. a
3. d	7. d	10. c
4. a		

PART C

1. spore	6. black bread mold
2. sprout	7. zygote
3. Oomycetes	8. Oomycetes
4. haploid	9. club
5. chitin	10. true

CHAPTER 7

THE GREEN PLANTS

Taxonomists group all green plants in the kingdom Plantae. This kingdom consists of some single-celled species which include free-living cells,
cells that live together in colonies and some cells that adhere together
in long filaments. However, most members of the Plantae are multicellular. Although multicellular means "having many cells," the concept of
multicellularity involves more than numbers of cells. It embraces several
ideas about cell structure and function. One such idea concerns *cell
specialization* in which cells are "programmed" to carry out special tasks.
Cell specialization in multicellular plants also brings with it a *division of
labor* in which groups of cells in tissue formation work together to perform
some special life function of benefit to the entire plant organism. Another
condition necessary to all specialization is the evolution of cell structures
that permit specialization and difference in cell function.

As implied in the term "green plants," members of the kingdom Plantae contain the green pigment chlorophyll. Not only does chlorophyll
color plant leaves and some stems green, it, more importantly, traps light
energy which is used in the process of photosynthesis. As an outcome
of photosynthesis nutrient molecules are made, serving as food for both
plants and animals.

Most species of the kindgom Plantae are nonmotile, anchored to one
place, and unable to move about, but a few of the lower plants are motile
for at least part of the life cycle. However, the evolutionary trend exhibited in green plants is toward stationary organisms that carry out their
life functions on land in locations where they remain for life. Lower plant
species equipped to swim about live in salt and fresh water. Higher plants
are *terrestrial* (land-dwelling).

THE PLANT KINGDOM

Traditional taxonomists employ the technique of dividing the kingdom Plantae into two major groups: the Thallophyta and the Embryophyta. This type of division permits the convenience of comparing the structures of simple plants with the more complex tissues and organs of higher plants.

THALLOPHYTA

The *thallophytes* are not differentiated into roots, stems and leaves. The simple body of the plant—either a single cell or a flat sheet of simple cells—is known as a *thallus*. Thallophytes do not have specialized tissues to carry water, to anchor the plants, or to grow new cells. As a rule, the sex cells of the thallophytes are produced in rather simple sex organs that are not protected by a surrounding wall of cells. The zygotes of the thallophytes do not develop into embryos that are contained in a female reproductive organ. Most species of thallophytes live in fresh water, although there are a few saltwater forms. Some species live in damp soil or on the bark of trees.

The green algae, the brown algae and the red algae compose the three thallophyte divisions.

Chlorophyta—Green Algae

Green algae belong to the Chlorophyta. The scientific name tells us that these algae contain the green pigment chlorophyll. Of interest to botanists is the fact that members of the Chlorophyta are considered to be the ancestors of the higher land plants based on three points of evidence. First of all, green algae have chlorophyll *a* and *b* in the same amounts as cells in higher plants. Second, green algae store food in the same form as do the higher plants. Third, green algae, like the higher plants, have well-defined cell walls made of cellulose.

The Chlorophyta includes forms such as *Chlamydomonas* that typically live as single cells, simple and complex colonial forms, and forms like sea lettuce that typically undergo a life cycle known as alternation of generation.

Chlamydomonas

Chlamydomonas is a unicellular autotroph that is a representative type of green algae. In some ways it resembles cells of the protist kingdom, because it lives in water, is motile, and has a pair of flagella (Fig. 7.1). However, biochemical analysis of its chlorophylls, carotenoids, and stored food reveals that these compounds are identical to those in the cells of green land plants. Scientists believe that *Chlamydomonas* represents the general type of unicellular algae from which multicellular green plants evolved.

Chlamydomonas is an oval-shaped cell with a haploid nucleus. Although its pigments and starch granules are identical with those of higher plants, its cell wall, unlike that of other green algae, is composed of glycoprotein, not cellulose. Sticking out from the anterior end of the

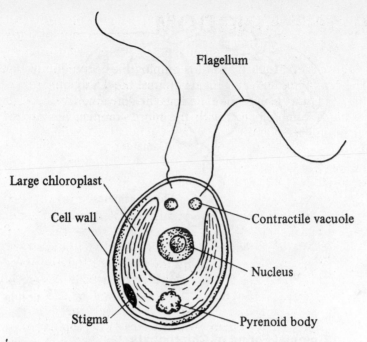

Fig. 7.1 *Chlamydomonas*

organism are two flagella of equal length. The large cup-shaped chloroplast fills up nearly the entire volume of the cell. Inside the chloroplast there are membraneous sacs that contain the chlorophyll. At the bottom portion of the chloroplast is a rather large circular *pyrenoid body* that functions in the synthesis and storage of starch. The *stigma*, a light sensitive eyespot, is located inside the chloroplast. Two contractile vacuoles which open and close alternately lie at the base of the flagella.

Before the onset of cell division, certain changes take place in the vegetative *Chlamydomonas* cell. The organism attaches itself to some object in the water and both of the flagella are resorbed into the cytoplasm. The once motile cell is now sessile. Mitosis and cytokinesis take place resulting in two identical daughter cells. These cells are retained within the membrane of the parent cell and quickly divide once more. Four identical flagellated *zoospores* are now released from the confines of the old cell wall. These cells grow and mature into vegetative *Chlamydomonas* organisms. This form of reproduction is asexual, and it is the usual mode of reproduction in *Chlamydomonas*.

Under unfavorable conditions (low nitrogen content in the water, for example), *Chlamydomonas* undergoes a primitive form of sexual reproduction. The vegetative *Chlamydomonas* cells go through several mitotic divisions, releasing several small flagellated cells into the water. These cells are known as *isogametes*, gametes that are indistinguishable from each other. The isogametes pair off, attached to each other end to end by the flagella. The cell wall slips away from each cell. Their cytoplasms fuse, forming a diploid zygote. A thick protective cell wall surrounds the zygote before it falls to the bottom of the pond. The zygote remains inactive and survives such an extreme condition as the drying of the pond. When environmental conditions improve, the zygote goes through a meiotic division forming zoospores that mature into haploid vegetative *Chlamydomonas* cells (Fig. 7.2).

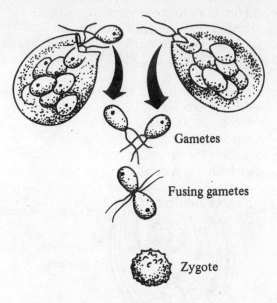

Fig. 7.2 Gamete formation in *Chlamydomonas*

Colonial Forms of Chlorophyta

The evolutionary trend from the unicellular *Chlamydomonas* to cells that can survive only when associated in a colony can be traced quite efficiently in the Chlorophyta. *Gonium* is a genus in the division Chlorophyta. Each *Gonium* species lives as a simple colony of cells. The cells of *Gonium* look very much like the single cell of *Chlamydomonas*. However, instead of living singly, they are held together in a gelatinous disk. The colony swims as a unit and the cells divide at the same time. Sexual reproduction in *Gonium* follows the pattern set by *Chlamydomonas* in that isogametes fuse to form a zygote.

    Pandorina, another colonial form of green algae, shows a bit more complexity than *Gonium*. In *Pandorina*, the colony has an anterior end and a posterior end as shown by the orientation in swimming. Cells in the *Pandorina* colony cannot live alone and the colony dies if broken apart. The male gametes of *Pandorina* are smaller than the female gametes. Since the gametes can be distinguished, this type of sexual reproduction is known as *heterogamy*. Since their only distinguishing characteristic is size, they are known as *anisogametes* (Fig. 7.3).

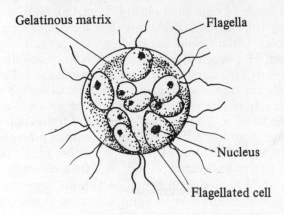

Fig. 7.3 *Pandorina*. The cells are confined in an intercellular bed or matrix of gelatinous material.

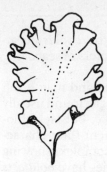

Fig. 7.4 Sea lettuce

Further colonial development is shown in the genus *Eudorina*. The cells of the anterior portion are more dominant than those of the posterior region, and the nonmotile female gametes are fertilized by the smaller, free-swimming male gametes inside the colony. The condition in which the sperm cell is motile and the egg cell is nonflagellated and nonmotile is called *oogamy*.

Colony development in *Volvox* is significantly more advanced than in any of the other green algae. Colonies in the *Volvox* genus are large, consisting of 500 to 50,000 cells. Cytoplasmic strands between the cells permit communication. Most of the cells are vegetative and do not function in reproduction. Scattered in the posterior half of the colony are a few large cells which are specialized for reproduction. Among these colonial forms we see an increase in the number of cells that make up the colonies, in the communication among cells, and in the specialization and differentiation of cells.

Alternation of Generations

The sea lettuce *Ulva* is a green algae that lives in saltwater. Its body is in the form of a flat leafy thallus and is two cell layers thick (Fig. 7.4). The life cycle of *Ulva* is described as *alternation of generations* because one generation of *Ulva* is produced sexually by gametes while the next generation is produced asexually by zoospores. To the naked eye, there is no difference in the structure of the sea lettuce produced sexually or asexually.

The *gametophyte generation* is a haploid thallus from which small, flagellated gametes are released into the water. They pair off and fuse. Each fused pair of gametes forms a zygote that, after a short time, becomes a diploid thallus of a new generation of *Ulva*.

The diploid thallus is the *sporophyte generation*. It produces haploid zoospores that develop and grow into a haploid thallus which is now the gametophyte (Fig. 7.5).

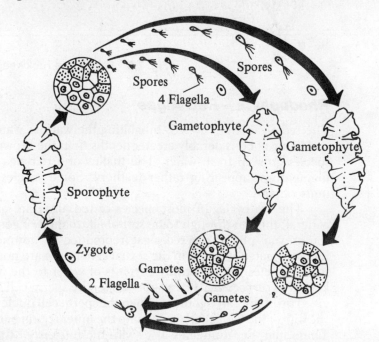

Fig. 7.5 Alteration of generations in sea lettuce (*Ulva*)

Phaeophyta—Brown Algae

Most species of brown algae are marine, inhabiting the cooler ocean waters. They grow attached to rocks along the shallow waters of seacoasts. The brown algae are commonly called seaweed.

Species of the Phaeophyta are the largest of the algae and may reach lengths of 45 meters or more. *Kelps* are massive brown algae found usually on the Pacific Coast. They contain chlorophylls *a* and *c*, but lack chlorophyll *b*. The brownish pigment *fucoxanthin* gives them their characteristic color. In structure, the brown algae are quite complex, showing considerable differentiation among the cells. Most species have *holdfasts* which secure them to rock substrates. Some species have stem-like and leaf-like parts. Many species have *air bladders* which give them buoyancy.

Fucus is a representative species of the Phaeophyta. Fig. 7.6 shows its general structure, including the very prominent air bladders. *Fucus*, also known as bladder wrack and rock weed, lives on the rocky sea shores of temperate seas. The thallus of *Fucus* is unusual in that it has repeated double branches, with enlarged tips commonly called *conceptacles*. These conceptacles hold the sex organs. The *antheridia* contain the sperm, while the *oogonia* hold the egg cells. In some species the male and female sex organs are on separate plants; in other species they are produced in the same conceptacle. Gametes are released into the water through pores at the tips of the conceptacles. An egg cell fuses with a sperm cell to form a zygote. The zygote grows into a diploid plant. The haploid gametophyte stage is completely missing in *Fucus*.

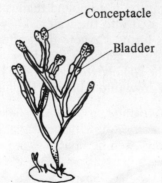

Fig. 7.6 *Fucus*, the rockweed

Rhodophyta—Red Algae

The red algae are seaweeds inhabiting the warmer waters of the oceans and living at considerably greater depths than the brown algae. Very few species grow in fresh water. The thallus of a red alga is branched and filamentous, appearing rather feathery. Some species resemble gelatinous ribbons.

The life cycles of most species of red algae are quite complex. In general, they go through some sort of alternation of generations. During the asexual phase, the red algae reproduce by nonmotile spores. The heterogametes involved in the sexual generation are nonmotile also. The male gametes are carried by currents of water to the female sex organs known as *carpogonia*. For a short while, the sperm cell sticks to the elongated tip of the carpogonium. Then the sperm cell nucleus passes through the wall of the elongated tip and into the inner region of the carpogonium. There, the sperm nucleus fuses with the nucleus of the egg cell.

Some representative species of red algae are *Chondrus* (also called Irish "moss"), *Polysiphonia* and *Nemalion*. The cell walls of the red algae are made of a combination of cellulose and a gelatinous material. The reserve carbohydrate stored in the cells of red algae is known as *floridean starch*; it is not a true starch.

The Rhodophyta contain both chlorophyll *a* and chlorophyll *d*. The latter is a type of chlorophyll not found in any other species of plants. The characteristic color of the red algae is given by the pigment *phycoerythrin*.

The value of accessory pigments to the photosynthetic process is demonstrated by the red algae. Chlorophyll *a* is the pigment actively involved in trapping light energy for use in photosynthesis. However, chlorophyll *a* cannot trap light energy at the depths at which the red algae grow. The pigments *phycocyanins* and phycoerythrins can and they pass the energy along to chlorophyll *a*.

Red algae have a great deal of commercial value. They are dried and ground up to be used as agar for bacterial media. They also form colloids which are used as suspending materials in ice cream and binders in puddings and chocolate milk.

EMBRYOPHYTA

The Embryophyta are land plants. They have developed a number of adaptations for life on land—structures and biochemical methods for conserving water which is necessary for all of the biochemical activities of the cell and thus for the maintenance of life. Chief among these adaptations was the evolution of sex organs that are protected by a surrounding layer of nonreproductive cells. *Antheridia* are male sex organs where sperm cells are produced. Egg cells are contained in organs called *archegonia*. Gametes enclosed in these organs are protected against drying out. Another adaptation for the prevention of water loss in the embryophytes is the *cutin* covering, a waxy substance impregnated in cell walls, which provides waterproofing to epidermal tissue and prevents evaporation of water. The higher embryophyta also have special *vascular* or water-carrying tissues designed to distribute water efficiently throughout the plant body.

Bryophyta—Mosses, Liverworts and Hornworts

The bryophytes are the first green land plants. They are primitive, small and inconspicuous. Although multicellular, the tissue differentiation is quite simple. Bryophyte species have no tissues that are specialized for water-carrying and no cambium specialized for growing new cells. Bryophyte species do not have true stems, leaves or roots. Simple rootlike structures called *rhizoids* anchor the plants to the ground and absorb moisture from the soil.

Alternation of generations occurs in all bryophytes. The larger and more noticeable generation is the gametophyte which usually supports a smaller (sometimes parasitic) sporophyte. Ciliated sperm cells, produced in the antheridia, swim to the archegonia and fertilize the egg cells held therein. The sporophyte generation begins with the zygote and is therefore diploid. Special cells in the sporophyte go through meiosis and produce haploid spores that begin the gametophyte generation.

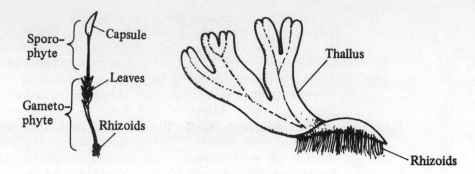

Fig. 7.7 Moss   **Fig. 7.8** Liverwort

There are 23,000 species of bryophytes among which are the mosses (Fig. 7.7), the liverworts (Fig. 7.8) and the hornworts, a group of plants structurally similar to the liverworts.

Tracheophyta—Vascular Plants

The vascular plants are truly land-dwelling plants. They have developed adaptations that permit them to live on land independent of bodies of water. The word "vascular" means that these plants have a water-carrying system. Water is conducted upward from the roots by *xylem* tubules. Fluid compounds are conducted downward from the leaves to lower plant organs by the *phloem* tubules.

The tracheophytes are divided into five subdivisions: psilopsids, club mosses, horsetails, ferns and seed plants.

Psilopsida

There is disagreement among botanists as to whether there are two living genera—*Psilotum* and *Tmesipterteris*—of psilopsids or whether all members of this subdivision are extinct. The plants that may represent the psilopsids have rather simple bodies, with branched stems, but no roots. The leaves are absent or very small.

Lycopsida—Club Mosses

Fig. 7.9 Club moss—*Lycopodium*

There are about 900 living species of club mosses. Plants in this group are usually one meter or less in height. Many are ground creepers. The club mosses have water-carrying (vascular) tissues and true roots, stems and leaves. The leaves are small and spirally arranged on the stems. The popular name of the Lycopsida is derived from the arrangement of the sporangia which are clustered on leaves formed into *cones* or *strobili*. The cones are positioned on the tips of stems as shown in Fig. 7.9. Some club moss species bear one type of spore and one type of gametophyte; such species are described as being *homosporous*. Other species of club moss are *heterosporous* because they bear two types of spores and produce two types of gametophytes.

Sphenopsida—Horsetails

Only 25 living species of Sphenopsida remain. The horsetails are true land plants having a vascular system, true roots, stems and leaves. Although the leaves are small and scale-like, they carry on photosynthesis. Horsetails grow well in both tropical and temperate climates, along river banks and in moist tracts of land (Fig. 7.10).

The life cycle of the horsetails involves alternation of generations. The sporophyte generation is the conspicuous generation. Its haploid spores give rise to a small plant known as a *prothallus*. The prothallus represents the gametophyte generation. The zygote remains attached to the gametophyte and ultimately becomes the sporophyte.

The cell walls in the leaves and stems of horsetails contain silica (sand), making the dried, ground stems useful as scouring powders.

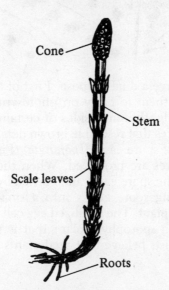

Fig. 7.10 The horsetail—
Equisetum

Pteropsida—Ferns

The fern plant used in flower bouquets is the sporophyte generation. Remember that the sporophyte generation produces asexual spores. The mature fern has true roots, leaves and stem. Ferns growing in temperate climates have an underground stem called a *rhizome* which grows in a horizontal position. The rhizome not only stores food materials, but also gives rise to new fern plants which grow along its length. The stems of tropical species grow upright in a vertical position and serve as trunks of tree ferns.

Fern stems are composed of xylem and phloem vascular tissues but they do not have the growth layer of embryonic cells known as *cambium*. Growing from the lower side of the rhizome are fibrous roots which absorb water and dissolved minerals from the soil. The leaves grow out from the top of the rhizome and break through the ground. Ferns have compound leaves composed of many leaflets.

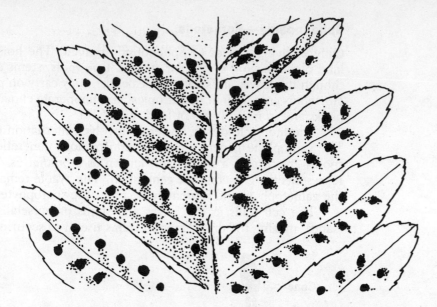

Fig. 7.11 Fern leaf showing spore cases

The leaves of ferns serve a dual purpose. First of all, the green pigment chlorophyll enables them to carry on photosynthesis to produce food for the plant. Secondly, the undersides of certain green leaves are covered with small structures that resemble brown dots. Each dot, called a *sorus*, is really a cluster of spore cases (*sporangia*) (Fig. 7.11). Inside of the sporangia haploid spores are produced. When the spores are ripe, they are discharged from the spore cases and carried by the wind to new soil habitats. A germinating spore grows into a small, inconspicuous heart-shaped gametophyte plant. The fertilized egg cell becomes a zygote that develops into a young sporophyte plant that is nourished by the gametophyte. Many botanists believe that seed plants evolved from the ferns.

Spermopsida—Seed Plants

The seed plants are divided into two groups: the *gymnosperms* and the *angiosperms*. The gymnosperms are known as the naked seed plants, while the angiosperms are referred to as the covered seed plants. Seed plants are the most successful land plants that have ever lived because they have reproductive mechanisms that do not require water. The plants that evolved before the seed plants produce sperm cells that can reach the egg cells only by swimming through a film of water.

Angiosperms are the most recently evolved major group of plants, dating from the early Cretaceous period, about 65 million years ago. Their success is measured in terms of the increase in number of species and their emergence as the dominant groups of plants, inhabiting varying environments throughout the world.

Gymnosperms Gymnosperms are cone-bearers. They are woody plants, chiefly evergreens, with needle-like or scale-like leaves. Cone-bearing plants grow in many parts of the world including tropical climates. However, most species are found in the cooler parts of temperate regions. Examples of gymnosperm species are pines, spruces, firs, cedars, yews, California redwoods, bald cypresses and Douglas firs. Most biologists do not think that the gymnosperms evolved directly from ferns.

Reproduction in gymnosperms takes place on special structures called *cones*. The cones are specialized nongreen leaves where seeds are produced. Most gymnosperm species bear two different kinds of cones: seed cones, called *megasporophylls*, and pollen cones, called *microsporophylls*.

Seed cones are large and their woody "leaves" (sporophylls) bear on them an *ovule* (called also a *megasporangium*). Inside of the ovule is a *megaspore mother cell*. This mother cell goes through a meiotic division producing both an egg cell and a food storage cell.

Pollen cones are smaller in size than the seed cones. Their leaves are known as *microsporophylls*. These microsporophylls are really *stamens*, reproductive structures that produce pollen. Inside of each pollen grain is a *microspore* cell. When the pollen grain germinates, the microspore goes through a reduction division producing two sperm cells.

Pollen grains are carried by the wind from the pollen cones to the seed cones. When a pollen grain lands on an ovule, it begins to germinate (sprout). The germinating pollen grain commences to grow a pollen tube during its first summer and the following spring. As the pollen tube grows, changes take place in the "leaves" of the seed cone where the ovules are pollinated. These leaves grow together holding the ovules tightly inside of the cone. The pollen tube enters a pore (*micropyle*) in the ovule. The two sperm nuclei from the pollen grain travel through the pollen tube into the ovule. One sperm cell fertilizes the egg cell. The other sperm cell and the pollen tube cells disintegrate. The fertilized egg becomes a zygote and then develops into the embryo plant. Changes occur in the outer walls of the ovule where a tough seed coat develops.

The ripened ovule is now a *seed* containing both an embryo plant and food for the embryo plant. The stored food known as *endosperm* is derived from the female gametophyte. The embryo plant represents the new sporophyte generation. Seeds are carried to new locations by wind, water or animals. The embryo plant inside of a seed can live for long periods in a resting state. The outer coat of the seed protects the embryo from extremes of temperature, drying, chemical corrosion and even burning. The germinating pollen grain and the developing ovule represent the gametophyte generation. The large conspicuous plant is the sporophyte generation.

Angiosperms—Flowering Plants   The reproductive structure of the angiosperm is the *flower* which encloses the male and female sex organs. Fig. 7.12 shows the parts of the flower. On the outside of the flower are the green leaflike *sepals*. The sepals protect the flower when in the bud stage. Collectively, sepals are known as the *calyx*. Just inside of the calyx are colored *petals*, showy and conspicuous in flowers that are pollinated by insects or birds. All of the petals in a flower are known as the *corolla*. Look at Fig. 7.12 and locate the stamens. Notice that the top portion is called the *anther* and the stalk, the *filament*. The stamens are the male reproductive structures; pollen grains are produced in the anther. Locate the *pistil* which is in the center of the flower. The top portion of the pistil is the *stigma*. The *style* is the long stalk that leads to the rounded portion of the pistil called the *ovary*. Inside of the ovary are many *ovules*. The pistil and its many parts compose the female portion of the flower. The pistil is really a modified sporophyll which in the flowering plant is called the *carpel*.

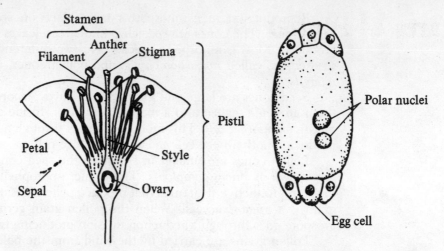

Fig. 7.12   Parts of a flower

Fig. 7.13   Maturation of the ovule

When an ovule is ready for fertilization, its megaspore mother cell goes through a reduction division and two mitotic divisions yielding eight nuclei. One of these nuclei forms the *embryo sac*. A nucleus near the ovule micropyle (pore) becomes the egg cell (Fig. 7.13).

Each pollen mother cell in the anther divides by meiosis and produces four pollen grains, each with a haploid nucleus. When a pollen grain lands on the stigma, its cell divides by mitosis to form two nuclei. One, the tube nucleus, directs the growth of the pollen tube down to the micropyle in an ovule. The other nucleus (*generative nucleus*) divides and forms two sperm nuclei (Fig. 7.14).

The sperm nuclei enter the ovule through the micropyle. One sperm fertilizes the egg cell and forms a diploid zygote. The other sperm cell fertilizes two polar nuclei in the embryo sac to form a triploid endosperm nucleus (Fig. 7.15). The behavior of both sperm cells is described as a *double fertilization*. The zygote then undergoes a number of changes and develops into the embryo plant. The ovule coats increase in number and harden, changing into seed coats. After fertilization in the ovules, the ovary enlarges and usually becomes a *fruit*. The ovules enlarge, change shape and form *seeds*, each containing an embryo plant and stored food. All accessory nuclei disintegrate.

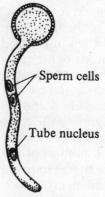

Fig. 7.14   Maturation of the pollen grain

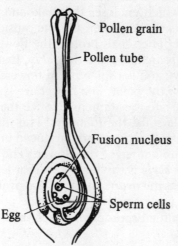

Fig. 7.15   Fertilization in the flower

Parts of Higher Plants

The bodies of the higher plants are differentiated into roots, stems, and leaves with specialized tissues to carry water, transport nutrients, anchor the plants, grow new cells, and carry out other functions of the plant.

ROOT

Root Function

Roots anchor plants to the soil and absorb water and dissolved minerals from the ground. The absorbed materials enter the root by way of root hairs which are one-cell extensions of the epidermis. From the root hairs, dissolved materials pass through the cortex, endodermis, and pericycle into xylem cells. The xylem cells conduct the dissolved materials upward. The root cortex serves in the storage of food and water. A small amount of storage occurs in the parenchyma cells of the stele. Cells in the tips of the roots are responsible for the growth in length of the root. Growth in diameter of the root is controlled by the cambium between the xylem tissue and the phloem tissue. Asexual reproduction (vegetative propagation) is brought about in some plants by adventitious buds present on roots.

Gross Structure

There are two types of root systems: fibrous roots and tap roots. The *fibrous root* has numerous slender main roots of equal size with many branch roots smaller in size. Examples of plants with fibrous roots are corn, wheat, grasses (Fig. 7.16). The *taproot* is the main root of the plant. It is longer and thicker than the smaller branch roots. Examples of taproots are carrots, beets, dandelions (Fig. 7.17).

Microscopic Structure

A longitudinal section of a young root shows four zones distinguished by the cell types within each region (Fig. 7.18).

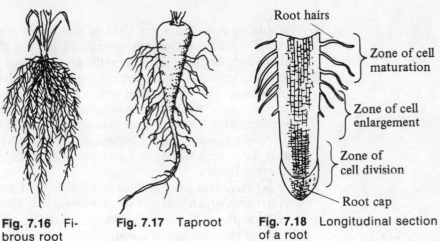

Root hairs

Zone of cell maturation

Zone of cell enlargement

Zone of cell division

Root cap

Fig. 7.16 Fibrous root　　**Fig. 7.17** Taproot　　**Fig. 7.18** Longitudinal section of a root

1. *Root cap.* A semicircular cap of cells forming the tip of the root and protecting the dividing cells just above it. Cells in the root cap are of moderate size and thick-walled. They protect the thinner-walled cells just above.

2. *Zone of cell division (meristemic region).* Here the cells are small, thin-walled with dense cytoplasm. These cells reproduce rapidly by mitosis and contribute to increased length of the root.

3. *Zone of cell enlargement or elongation.* The cells in this zone were recently formed in the zone of cell division. These cells become elongated, produce new cytoplasm and develop larger vacuoles.

4. *Zone of maturation.* In this zone the enlarged cells become *differentiated* into xylem, phloem, cambium, cortex and other tissues. Root hairs grow from the lower portion of the maturation zone.

A cross section of the root taken through the zone of maturation reveals the following tissues (Fig. 7.19):

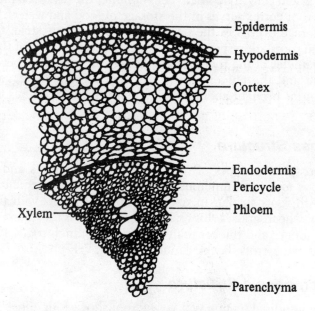

Fig. 7.19   Cross section of a root

1. The *epidermis* is the outer layer of cells from which the root hairs develop. Epidermal tissue is specialized for the absorption of water and minerals from the soil and for the protection of underlying tissues.

2. The *cortex* contains rather large, irregularly shaped parenchyma cells applied loosely to one another with a good deal of intercellular space. The function of this region is to store water and food. Water leakage from the cortex into the inner tissues is prevented by a band of thick-walled cells known as the *Casparian strip.* A waxy material, *suberin,* makes these cells waterproof and prevents leakage of water into the inner root tissues.

3. The *pericycle* is a layer of cells just inside the Casparian strip (endodermis) from which branch roots are produced. The cells of the branch roots grow through the cortex and through the epidermis and extend outside of the root into the soil.

4. The *xylem* is composed of conducting cells called *tracheids* and thin tubules known as *vessels*. The function of xylem tissue is to conduct water and dissolved substances upward through the root into the stem.
5. The *phloem* is composed of companion cells and sieve tubes. The function of the phloem is to transport water with dissolved food downward from the leaves through the stem into the root.
6. The *parenchyma* stores food and water and lends support to other tissues.

STEM

Stems have three major functions. First, they conduct water upward from the roots to the leaves and conduct dissolved food materials downward from the leaves to the roots. Second, stems produce and support leaves and flowers. Third, they provide the mechanisms for the storage of food.

Specialized stems have additional functions. The *tendrils* of grape vines function as climbing organs. *Thorns* on rose bushes offer protection against animal invaders. The desert cacti have stems specialized for the storage of water and food. *Runners* of strawberry plants and spider plants serve as organs of *vegetative propagation* in which new plants are produced at their nodes. *Rhizomes* are underground stems of ferns which serve the purpose of producing new plants. The underground stem of the white potato is called a *tuber* and its function is to store carbohydrate in the form of starch. The *corms* of the crocus and gladiolus are underground storage stems consisting of fleshy leaves.

Botanists call a stem with its leaves a *shoot*. The *shoot system* is the total of all of the stems, branches and leaves of a plant.

Fig. 7.20 shows a longitudinal section of a young stem.

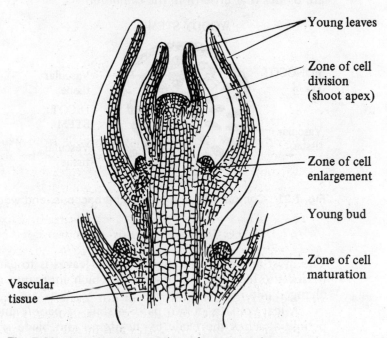

Fig. 7.20   Longitudinal section of a young stem

TABLE 7.1. Types of Aerial Stems

	Herbaceous	Woody Stems
Texture	soft	tough
Color	green	nongreen
Growth	little in diameter	much growth in diameter
Tissues	primary	secondary
Life cycle	annual	perennial
Protective covering	epidermis	bark
Buds	naked	covered with scales
Examples	monocotyledons	all gymnosperms
	dicotyledons	dicotyledons

There are two types of above ground (aerial) stems: woody stems and herbaceous stems. Major characteristics of each are shown in Table 7.1.

Herbaceous stems of dicotyledons have the following tissues: epidermis, schlerenchyma, cortical parenchyma, pericycle, phloem, cambium, xylem, and pith. Herbaceous stems of monocotyledons do not have cambium. Since no secondary growth takes place, the stems do not increase in size appreciably. Microscopic study of the stem cross section shows that the xylem and phloem are organized into vascular bundles. These are scattered throughout the parenchyma tissue which fills the stem.

Woody stems are composed of primary and secondary tissues. Primary tissues are those that develop from the meristems (embryonic tissue) of the buds on twigs during the first year of growth. After the first year, growth in the the woody stem takes place in the secondary tissues. These are tissues that arise from the *cambium.*

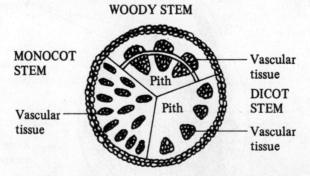

Fig. 7.21 Comparison of monocot, herbaceous, and woody stems

LEAF

The most important function of green leaves is to carry out photosynthesis, the food-making process in which inorganic raw materials are changed into organic nutrients.

A leaf consists of two parts: a stalk or *petiole* and the *blade*. The petiole attaches the blade to the stem. The blade is the place where photosynthesis takes place. Leaves vary greatly in shape (Fig. 7.22).

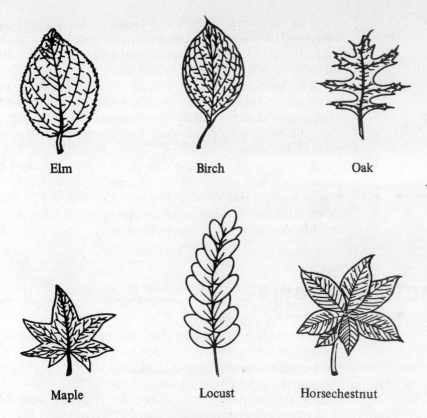

Fig. 7.22 Some common leaf shapes

Study of a leaf cross section under the microscope reveals three types of tissue: upper and lower epidermis, mesophyll, and the vascular bundles (Fig. 7.23).

The *epidermis* is a single layer of cells at the upper and lower surfaces of the leaf. The cells have thick walls made of cutin and lack chloroplasts. Their main function is to protect the underlying or overlying tissues from drying, bacterial invasion and from mechanical injury. On the underside of the leaf, the lower epidermis has pores known as *stomates*, the size openings of which are regulated by a pair of *guard cells*. The stomates serve as passageways for oxygen and carbon dioxide.

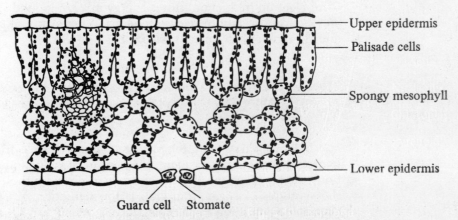

Fig. 7.23 Leaf cross section

The *mesophyll* is positioned between the two epidermal layers. It is composed of two distinct tissues: the *palisade cells* and the *spongy cells*. Packed with chloroplasts, the cells of both types of mesophyll tissue are the sites for photosynthesis in the leaf.

The *vascular bundles* or veins consist of xylem cells (vessels and tracheids) and phloem cells (sieve tubes and companion cells). The xylem cells transport water and dissolved minerals upward into the mesophyll. The phloem conducts dissolved food materials downward to the lower parts of the plant.

Loss of water vapor through plant leaves is termed *transpiration*, a process which is responsible for the rise of sap in trees. *Guttation* is the loss of liquid water through the leaves of plants with short stems. Guttation is caused by the effect of root pressure on water flow and has hardly any physiological advantage to the plant.

Photosynthesis

Photosynthesis takes place in the chloroplasts of green leaves and stems. Photosynthesis is the food-making process of green plants made possible by the light-trapping pigment chlorophyll. An important event of photosynthesis is the changing of light energy into chemical energy, which is ultimately stored in sugar molecules. The raw materials of photosynthesis are carbon dioxide and water. The overall equation for photosynthesis is shown below:

$$CO_2 + 2H_2O + light \xrightarrow{\text{chlorophyll}} O_2 + (CH_2O) + H_2O$$

REDUCTION-OXIDATION REACTIONS (REDOX)

Reduction is the addition of an electron (e) to an acceptor molecule. Oxidation is the removal of an electron from a molecule. The addition of an electron (reduction) stores energy in a compound. The removal of an electron (oxidation) releases energy. Whenever one substance is reduced, another substance is oxidized.

A^e	+	B	→	A	+	B^e
electron donor		electron acceptor		oxidized (loss of energy)		reduced (gain of energy)

In biological systems, removal or addition of an electron derived from hydrogen is the most frequent mechanism of reduction-oxidation reactions. Redox reactions play a major role in photosynthesis. For example: the synthesis of sugar from CO_2 is the reduction of CO_2. Hydrogen, obtained by splitting H_2O molecules, is added to CO_2 to form CH_2O units.

THE PROCESS OF PHOTOSYNTHESIS

Photosynthesis takes place inside of chloroplasts, membranous structures within the cells of the leaf mesophyll. The chloroplasts have fine structures within—flattened membranous sacs named *thylakoids*. On the membranes of the thylakoids, chlorophyll and the accessory pigments are organized into functional groups known as *photosystems*. Each of these photosystems contains about 300 pigment molecules which are involved directly or indirectly in the process of photosynthesis.

Each of these photosystems has a *reaction center* or a *light trap* where a special chlorophyll *a* molecule traps light energy. There are two types of photosystems: Photosystem I and Photosystem II. In Photosystem I, the chlorophyll *a* molecule is named P 700 because it absorbs light energy from the 700 nanometer wavelength. The chlorophyll molecule of Photosystem II is designated as P 680 because this pigment molecule (chlorophyll *a* absorbs light at the wavelength of 680 nanometers.

Photosynthesis involves four sets of biochemical events: photochemical reactions, electron transport, chemiosmosis and carbon fixation. The photochemical reactions and electron transport activities take place on the membranes of the thylakoids. The oval membranes of a thylakoid surround a vacuole or reservoir in which hydrogen ions are stored until needed in the Calvin cycle, or carbon fixation. Each thylakoid rests in the *stroma* or ground substance of the chloroplast. The stroma is the place where carbon fixation occurs.

The Events of Photosystem I

Cyclic Phosphorylation—Electron Transport

Light energy strikes a photosystem. Pigment molecules absorb this energy and pass it on to the reaction center molecule. The energy level of an electron in P700 is raised to a higher level. This increased energy in the electron causes it to escape the confines of the P700 (chlorophyll *a*) molecule and to become temporarily attached to an acceptor molecule called X. In accepting the electron, molecule X is reduced. Molecule X passes the electron on to another acceptor molecule and becomes oxidized in the process. In a series of redox reactions, the electron is passed from one acceptor molecule to another. It finally returns to P700. Each step of these reduction-oxidation reactions is catalyzed by a specific enzyme.

The energy released as the electron is passed along the transport chain is used to synthesize ATP (adenosine triphosphate). Excess hydrogen ions released when ATP is formed are stored in the reservoir of the thylakoid. Inorganic phosphate from the fluid of the stroma is incorporated in the ATP molecule during photosynthetic phosphorylation. Photosynthesis requires the energy from ATP to synthesize carbohydrates.

Noncyclic Phosphorylation

During the events of Photosystem I, "excited" electrons may travel in another pathway different from that which builds ATP. Chlorophyll acts as an electron donor and later as an electron acceptor. It donates energy-rich (excited) electrons and accepts back energy-poor electrons.

Light energy strikes a chlorophyll *a* molecule. An electron in its reaction center molecule, P700, becomes elevated to a higher energy state. The electron passes from P700 to acceptor X. From acceptor X, the electron passes to ferridoxin (Fd), an iron containing compound. Fd passes the electron to an intermediate compound and then to nicotinamide adenine dinucleotide (NADP). Actually two P700 molecules release electrons along this pathway simultaneously. NADP accepts both electrons (2e) and becomes NADPH. NADPH keeps both electrons and does not pass them along. The energy from $NADPH_2$ will serve as an energy source when carbon dioxide is reduced to form sugar. By acquiring two extra electrons, the NADPH also attracts a H proton. Thus NADPH + H is written and $NADP_{re}$.

Review: Photosystem I

1. Photons of light strike a chlorophyll *a* molecule.
2. The reactioin center molecule (P700) absorbs the light.
3. One of its electrons is raised to a higher energy level.
4. The pathways followed by "excited" electrons are shown below.

Cyclic   *e* from P700 to X to acceptors to ATP

Noncyclic   2e from P700 to X to Fd to NADP → $NADP_{re}$

Photosystem II

Photosystem II involves about 200 molecules in the reaction center of chlorophyll *a*, the light trapping pigment of green plants. In the blue-greens and in the bryophytes, the light trapping pigment is chlorophyll *b*; in the brown algae chlorophyll *c* and in the red algae chlorophyll *d*.

When light strikes the chlorophyll in Photosystem II, an electron in the reaction center molecule, P680, becomes "excited". The energized electron is passed to an electron "acceptor" molecule designated *Q*. Molecule *Q* passes the electron through a chain of acceptor molecules which pass the electrons to the holes in Photosystem I formed during the noncyclic synthesis of $NADP_{re}$. As electrons move along the chain of transport, they lose energy step by step. Some of the energy forms ATP. It is believed that P680 pulls replacement electrons from water, leaving behind free protons and molecular oxygen:

$$2H_2O \rightarrow 4e + 4H^+ + O_2$$
to P680

The protons become associated with $NADP_{re}$.

A summary equation reviewing the electron pathways in Photosystem I and in Photosystem II follows:

$H_2O \rightarrow 2e$ to Photo I to *Q* to transport chain to Photo II to X to transport chain to $NADP_{re}$ to Calvin Cycle.

Fig. 7.24 shows the pathways of photosynthesis.

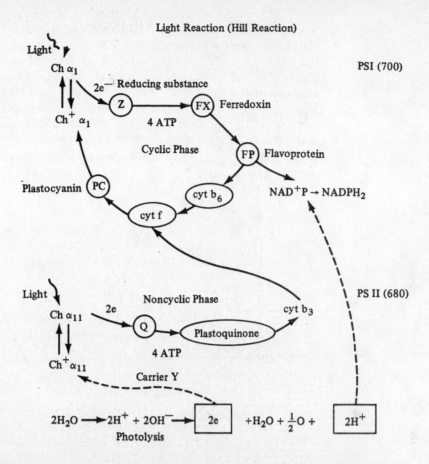

Fig. 7.24   The pathways of photosynthesis involve many steps and many intermediate products and catalysts, including flavoprotein and cytochrome (cyt).

The Calvin Cycle

The Calvin cycle is the series of events in photosynthesis during which carbon dioxide fixation takes place in the stroma of the chloroplast. $NADP_{re}$ and ATP that were produced during Photosystem I and Photosystem II are now used to attach CO_2 to a preexisting organic molecule. The enzymes that catalyze the events of the Calvin cycle are present in the stroma.

Carbon dioxide combines with the 5-carbon sugar, ribulose biphosphate (RuBp) forming an unstable 6-carbon compound. This compound breaks into two 3-carbon compounds, phosphoglyceric acid (PGA). The two PGA molecules are reduced to two molecules of phosphoglyceraldehyde (PGAL) in two successive steps. First, the PGA molecules receive a high energy phosphate from ATP. The high energy phosphate bond is broken, the phosphate is removed and replaced by a hydrogen atom from NADPH. Then, the joining together of two PGAL molecules results in the formation of a molecule of 6-carbon sugar. Some of the PGAL is used to replenish the store of ribulose biphosphate, the starting point of the Calvin cycle (Fig. 7.25).

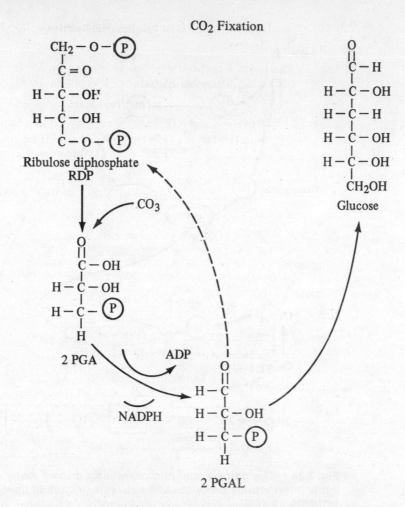

Fig. 7.25   The Calvin cycle showing the complex steps leading from ribulose diphosphate (biphosphate) to glucose, a 6-carbon sugar

Chronology of Famous Names in Biology

1679   **Marcello Malpighi** (Italian)—determined the functions of xylem and phloem by conducting a series of girdling experiments.

1729   **Stephen Hales** (English)—demonstrated that transpiration can pull sap up through the xylem and phloem.

1772   **Joseph Priestley** (English)—demonstrated that the air is replenished by green plants.

1782   **Jean Senebier** (French)—showed that photosynthesis depends on "fixed air," now known as carbon dioxide.

1796   **Jan Ingen-Housz** (Dutch)—concluded that plants use carbon dioxide in photosynthesis.

1804   **Nicolas Theodore De Saussure** (French)—demonstrated that water is necessary for photosynthesis.

1879	**James Clerk Maxwell** (English)—discovered that light travels in waves.
1883	**T. W. Engelmann** (Germany)—provided evidence that chlorophyll plays a role in photosynthesis and demonstrated that red light was most effective in photosynthesis.
1904	**B. Haberland** (German)—discovered plants native to tropical climates have a different arrangement of the bundle sheath cells.
1905	**F. F. Blackman** (English)—first to present evidence that photosynthesis has a light-driven stage and a stage not requiring light.
1905	**Albert Einstein** (American)—proposed that light energy travels in packets called photons.
1910	**Max Planck** (German)—established that the energy of radiation is contained in packets called quanta.
1934	**C. B. van Niel** (American)—proposed that water is the source of the oxygen in photosynthesis.
1941	**Samuel Ruben, Merle Randall, Martin Kamen** and **James L. Hyde** (American)—reported that oxygen liberated in photosynthesis comes from water.
1950's	**George Wald** (American)—greatest living expert on light and life.
1950	**Daniel Arnon** (American)—identified the reactions that take place in light reactions of photosynthesis.
1954	**Hugo Kortschak** (American)—discovered that Kranz anatomy plants begin carbon dioxide fixation with a 4-carbon compound.
1960	**H. P. Kortschak** (American), **M. D. Hatch** and **C. R. Slack** (both Australian)—elucidated photosynthetic pathways in C_4 plants such as sugarcane, corn and sorghum.
1961	**Melvin Calvin** (American)—discovered the pathway of events by which green plants incorporate carbon dioxide into carbohydrate molecules.

Words for Study

alternation of generations	carpogonia	embryophytes
anther	conceptacle	endosperm
antheridia	cone	fucoxanthin
archegonia	corolla	gametophyte
calyx	cutin	heterogamy
cambium	division of labor	holdfasts
carpel	embryo sac	isogametes

micropyle   pistil   specialization
microsporophyll   plantae   sporophyte
microspore   polar bodies   stamen
multicellular   pollen   stigma
oogamy   prothallus   strobilus
oogonia   pyrenoid body   thallophytes
ovary   rhizoids   thallus
ovule   rhizome   vascular
petal   sepal   xylem
phloem

Questions for Review

PART A.   Completion. Write in the word that correctly completes each
statement.

1. Single-cell plants include free-living forms, cells that live in colonies,
    and ..1.. forms.
2. The chromsome number of vegetative *Chlamydomonas* cells is best
    described as being ..2..
3. Primitive gametes that are similar in appearance are known as ..3..
4. *Gonium* species live as a simple ..4.. of cells.
5. Anisogametes can be distinguished by differences in ..5..
6. The *Volvox* series of cells demonstrates an evolutionary trend toward
    ..6..
7. Antheridia are best associated with ..7.. gametes.
8. Carpogonia are best associated with ..8.. sex cells.
9. Chlorophyll ..9.. is found only in the red algae.
10. The water-carrying tissues of green plants are known collectively as
    ..10.. tissues.
11. The cycle of a sexual generation following an asexual generation is
    known as ..11..
12. Fluid compounds are conducted downward in plants by ..12.. tub-
    ules.
13. A sorous (sori, pl) can be found on the leaves of ..13.. plants.
14. The dominant generation in the ferns is the ..14.. generation.
15. The gymnosperms are best described as ..15.. bearers.
16. The prefix "mega" means ..16..
17. The microsporophylls are ..17.. cones.
18. In gymnosperms, the agent of pollination is ..18..
19. In gymnosperms, the megasporangium produces the ..19..
20. Petals are known collectively as the ..20..

21. The gymnosperms (did or did not) evolve directly from the ferns.
22. The tube nucleus develops from the germinating ..22..
23. A fruit is a ripened ..23..
24. Growth in length of roots occurs at the ..24..
25. A stem with its leaves is called a ..25..
26. The Casparian strip is located in what part of the plant?
27. Secondary plant tissues arise from the ..27..
28. The most effective light-trapping pigment is ..28..
29. Carotene and phycoerythrin function as ..29.. pigments during the light driven stages of photosynthesis.
30. Carbon fixation takes place in the ..30.. of the chloroplat.

PART B.   Multiple Choice. Circle the letter of the item that correctly completes each statement.

1. Thallophytes do not have structures which permit
   (a) photosynthesis   (c) water-carrying
   (b) zygote formation   (d) reproduction

2. *Chlamydomonas* resembles the protists because it lives in water and swims by means of
   (a) cilia   (c) tentacles
   (b) pseudopods   (d) flagella

3. The function of the pyrenoid body is to
   (a) synthesize glucose   (c) resorb flagella
   (b) store starch   (d) secrete cellulose

4. Sea lettuce is best classified as a (an)
   (a) protist   (c) euglenoid
   (b) bryophyte   (d) alga

5. A true statement about volvox is
   (a) Most of its cells do not function in reproduction.
   (b) All of its cells are specialized for reproduction.
   (c) There is no communication between its cells.
   (d) The egg cells have paired flagella.

6. The sporophyte generation of sea lettuce is
   (a) haploid   (c) diploid
   (b) monoploid   (d) tetraploid

7. Fucus does not have
   (a) air bladders   (c) a sexual cycle
   (b) an asexual cycle   (d) a branching thallus

8. Accessory pigments are those that pass light energy to
   (a) carotene   (c) chlorophyll *a*
   (b) phyoerythrin   (d) chlorophyll *d*

9. The greatest hazard to land-dwelling plants is
   (a) drying out   (c) loss of sperm cells
   (b) insect infestation   (d) nonmotile egg cells

10. Waterproofing of land cells is made possible by cell walls composed of
    - (a) chitin
    - (b) cutin
    - (c) cellulose
    - (d) glycoprotein

11. Examples of bryophytes are
    - (a) moss, *Volvox*, *Fucus*
    - (b) Ulva, hornwort, *Pandorina*
    - (c) *Euglena*, moss, red algae
    - (d) hornwort, liverwort, moss

12. Species that bear two types of spores are described as being
    - (a) homosporous
    - (b) heterosporous
    - (c) homozygous
    - (d) heterozygous

13. Dried, ground stems of the horsetails are used in scouring powder because their cell walls contain
    - (a) cutin
    - (b) cellulose
    - (c) silica
    - (d) suberin

14. An underground stem is called a
    - (a) Prothallus
    - (b) root
    - (c) strobilus
    - (d) rhizome

15. Ferns are successful land plants because they have well developed
    - (a) vascular systems
    - (b) digestive systems
    - (c) reproductive systems
    - (d) excretory systems

16. In the gymnosperms, the megasporophyll is a (an)
    - (a) large spore
    - (b) seed cone
    - (c) egg cell
    - (d) large ovule

17. The embryo plant is protected inside of the
    - (a) ovule
    - (b) egg
    - (c) pollen
    - (d) seed

18 The production of two types of gametes is known as
    - (a) heterospory
    - (b) heterogamy
    - (c) isogamy
    - (d) isospory

19. A true statement about the gymnosperms is that they do not produce
    - (a) flagellated sperm
    - (b) seed cones
    - (c) nonmotile eggs
    - (d) pollen cones

20. In the flowering plants, the male reproductive structures are the
    - (a) calyx
    - (b) corolla
    - (c) pistils
    - (d) stamens

21. The chromosome number of the fertilized endosperm nucleus is best described as
    - (a) haploid
    - (b) monoploid
    - (c) diploid
    - (d) triploid

22. The part of the young root that pushes its way through the soil is the
    - (a) root hair
    - (b) tap root
    - (c) root cap
    - (d) root zone

23. The word "meristem" refers to
    - (a) rapidly growing cells
    - (b) old cells
    - (c) cells specialized for reproduction
    - (d) differentiated cells.

24. The primary function of the root cortex is to
    (a) absorb minerals from the soil
    (b) differentiate into other cells
    (c) store water and food
    (d) give rise to sperm cells

25. Secondary tissues in roots arise from the
    (a) meristem   (c) cambium
    (b) embryo   (d) phloem

26. A tendril is a specialized
    (a) stem   (c) flower
    (b) root   (d) leaf

27. Sap rises in trees due to the force created by
    (a) evaporation   (c) transpiration
    (b) transportation   (d) guttation

28. Photosystems are functional pigment groups located on the
    (a) proteins of the plasma membrane
    (b) membranes of the thylakoids
    (c) in the stroma of the chloroplasts
    (d) in fluids of vacuoles

29. As an outcome of cyclic phosphorylation
    (a) P700 is destroyed   (c) NADPH is synthesized
    (b) ATP is formed   (d) carbon is fixed

30. Ribulose biphosphate
    (a) begins the Calvin cycle   (c) breaks into two equal parts
    (b) functions as an enzyme   (d) is a 6-carbon sugar

PART C.   Modified True-False. If a statement is true, write "true" for your answer. If a statement is incorrect, change the under-lined word to one that will make the statement true.

1. Most species of thallophytes live in salt water.

2. *Chlamydomonas* is classified as a protist.

3. Zoospores represent sexual reproduction.

4. If the original parent colony of *Gonium* contained 8 cells, each new colony produced will contain 32 cells.

5. Oogamy refers to conditions in which the egg cell is motile.

6. The sporophyte generation is produced by spores.

7. Brown algae are called seaweed.

8. Water is carried upward in plants by special tubules known as phloem.

9. A strobilus is a stem.

10. The fern plant is the gametophyte generation.

11. A germinating fern spore develops into a gametophyte plant.

12. The reproductive mechanisms of seed plants do not require gametes.

13. A megaspore mother cell is found inside of an egg.

14. Reproduction in the gymnosperms takes place on special structures called <u>stamens</u>.

15. In gymnosperms, the egg cell develops from the <u>microspore</u> cell.

16. The flowering plants are called <u>gymnosperms</u>.

17. Green leaves that protect flower buds are called <u>petals</u>.

18. The stigma and the style are best associated with the <u>filament</u>.

19. The ovules are found inside of the <u>ovary</u>.

20. A seed is a ripened <u>ovary</u>.

21. Growth in diameter of roots takes place in the <u>epidermis</u>.

22. Two types of root systems are fibrous roots and <u>root hairs</u>.

23. The root pericycle gives rise to <u>tap</u> roots.

24. The phloem is composed of companion cells and <u>sieve</u> tubes.

25. The underground stem of the white potato is a <u>rhizome</u>.

26. Runners of strawberry plants are specialized for <u>climbing</u>.

27. All gymnosperms have <u>herbaceous</u> stems.

28. The two types of cells that compose the leaf mesophyll are the spongy cells and <u>meristemic</u> cells.

29. Oxidation occurs by <u>loss</u> of an electron.

30. An "excited" electron escapes to a <u>lower</u> energy level.

Answers to Questions for Review

1. filamentous
2. haploid
3. isogametes
4. colony
5. size
6. multicellularity (specialization)
7. male
8. female
9. d
10. vascular
11. alternation of generations
12. phloem
13. fern
14. sporophyte
15. cone
16. large
17. pollen
18. wind
19. ovule
20. corolla
21. did not
22. pollen grains
23. ovary
24. tips
25. shoot
26. root
27. cambium
28. chlorophyll a
29. accessory
30. stroma

PART B

1. c	11. d	21. d
2. d	12. b	22. c
3. b	13. c	23. a
4. d	14. d	24. c
5. a	15. a	25. c
6. c	16. b	26. a
7. b	17. d	27. c
8. c	18. b	28. b
9. a	19. a	29. b
10. b	20. d	30. a

PART C

1. fresh water
2. green alga
3. asexual
4. 8
5. nonmotile
6. gametes
7. true
8. xylem
9. cone
10. sporophyte
11. true
12. water
13. ovule
14. cones
15. megaspore mother cell
16. angiosperms
17. sepals
18. pistil
19. true
20. ovule
21. cambium
22. taproots
23. branch
24. true
25. tuber
26. reproduction
27. woody
28. palisade
29. true
30. higher

CHAPTER 8

INVERTEBRATES: SPONGES TO STARFISH

All animals belong to the kingdom Animalia, a grouping of 29 phyla. Twenty-eight of these phyla include animals called invertebrates because they do not have a vertebral column, or true backbone; the 29th phylum includes the vertebrates, animals with a vertebral column. This chapter discusses the basic organization of the animal body, some general characteristics of invertebrates, and seven phyla of invertebrates.

ORGANIZATION OF THE ANIMAL BODY

The organization of the animal body is different from that of the plant and requires a different set of terms for accurate descriptions.

Symmetry

The animal body form is often described in terms of *symmetry*, the relative position of parts on opposite sides of a dividing line. Symmetry helps to define the degree of similarity between two species, or between parts of the same animal. *Spherical* symmetry describes the symmetry of a ball. Any section through the center of a ball-shaped organism divides the animal into equal and symmetrical halves (Fig. 8.1). *Radial symmetry* describes the symmetry of a wheel in which the parts are arranged in a circle around a central hub or axis. A vertical cut through the central axis divides a wheel-shaped organism into equal (symmetrical) halves (Fig. 8.2). Radial symmetry is a characteristic of animals that are sessile. However, radially symmetrical animals that have some head development are capable of rather quick movements. *Bilateral symmetry* is characteristic of animals that have a head end, a tail end, and right and left sides. The body of a dog or a horse is a good example of bilateral symmetry. A line drawn through the center of the body lengthwise divides the body into halves that are mirror images of each other (Fig. 8.3).

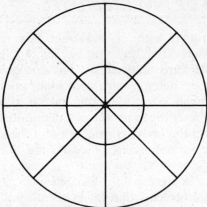

Fig. 8.1   Symmetry through a ball   **Fig. 8.2**   Radial symmetry

Fig. 8.3   Bilateral symmetry of the human body

Embryonic Cell Layers

All of the tissues, organs, and systems of animals develop from two or three *embryonic cell layers*, often referred to as *primary germ layers*. These three cell layers are the *ectoderm* (meaning outside skin), the *mesoderm* (middle skin), and the *endoderm* (inside skin). Animals with simple body structures may develop from only ectoderm and endoderm because the middle tissue layer remains undifferentiated. The tissues of animals with more complex body form develop from the three primary germ layers. In these animals there is a body cavity or *coelom* (see-lom) which is lined with endoderm.

Body Directions

The kinds of directions used to locate body structures permit precise description. The forward end of an animal is the *anterior* end. The *posterior* end is opposite to the anterior region. Example: the head end of a dog is the anterior end; its tail represents the posterior end. *Dorsal* refers to the back of an animal. In all animals except humans (and some primates) the dorsal side faces upward. The underside or bellyside of an animal is the *ventral* side. The sides of an animal such as the right hand and left hand side of a human or the right fore and hind limbs or left fore and hind limbs of a horse are the *lateral* sides of the body. The point of attachment of a structure, such as the wing of an insect, is the *proximal* end. The free, unattached end is known as the *distal* end.

GENERAL CHARACTERISTICS OF INVERTEBRATES

Invertebrates are animals without backbones. About ninety percent of all animal species are invertebrates. Just like all members of the kingdom Animalia, invertebrates are multicellular. They are composed of cells that lack walls. Animals are heterotrophs, dependent upon food supplied by the autotrophs. Animals ingest food; they take it in, digest it, and then by some means characteristic to the species, distribute nutrient molecules to cells that make up the body. Nutrients not utilized for energy or incorporated into the tissue-building processes are stored in cells as glycogen or fat.

Most invertebrates are capable of locomotion and have specialized cells with contractile proteins that facilitate movement. However, the adult forms of some lower invertebrate species are sessile, belonging to a group of *filter feeders*. These animals use cilia, flagella, tentacles or gills to sweep smaller organisms from the currents of water that flow over or through their bodies into the digestive cavities.

Some of the lower invertebrates reproduce vegetatively by *budding*. This means simply that a new organism grows from cells of the parent, breaks off, and then continues its own existence. Other invertebrate species reproduce sexually, utilizing sperm and egg. Still other species reproduce asexually by *parthenogenesis* in which an unfertilized egg develops into a complete individual. Some invertebrates have marvelous powers of *regeneration*, the growing back of lost parts or the production of a new individual from an aggregate of cells or from a piece broken off from the parent organism.

Phylum Porifera—Sponges

In a practical sense, you are familiar with the word *sponge*. You know that a sponge is able to soak up large amounts of water because of the many pores and spaces that form its structure. Living sponges are pore-bearing animals and, therefore, were given the scientific name of *Porifera*. These are the simplest of the multicellular animals. About 15,000 sponge species live attached to rocks bathed by ocean waters. Only a few species are freshwater dwellers.

STRUCTURE

Fig. 8.4 illustrates the structure of a simple sponge. Notice that the sponge has a cylindrical shape resembling a vase. At the anterior end is the opening of the central cavity which extends through the Porifera body. Sponges are composed of undifferentiated cells not organized into specialized tissues. The cells show a division of labor much like that of the cells in an algal colony. The cells of the sponge are arranged around the central cavity. In some species this cavity is divided into compartments; in others, it is a continuous space. The central cavity is brought into contact with the surrounding water by means of pores that lead into a system of canals.

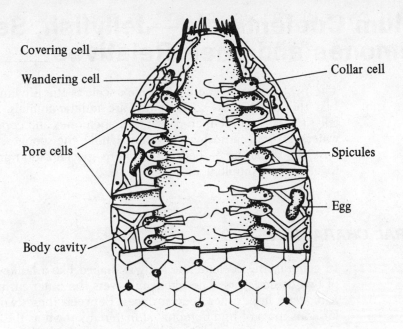

Covering cell

Wandering cell

Pore cells

Body cavity

Collar cell

Spicules

Egg

Fig. 8.4 Internal structure of the sponge

If you have ever held a natural sponge, you are aware that it is not flimsy. Sponges have an internal skeleton. In some species the skeleton is made of needle-like crystals called *spicules*, which are made of lime or silica (glass). In other species, a fibrous protein called *spongin* forms the skeleton, making the sponge firm, but pliable, wettable and absorbent.

Sponges are filter-feeders. The internal cavity is lined with *collar cells* equipped with flagella which capture microorganisms from water as it flows through the central cavity. Epithelial cells cover the outer body wall of the sponge. Between the outer wall and the cells that line the inner cavity is a jelly-like layer known as the *mesoglea*. Here amoeboid wandering cells known as *mesenchyme cells* help to transport digested molecules from the collar cells to cells in other parts of the body. The mesenchyme cells have the ability to change their form and function. They can change into collar cells, epithelial cells, or cells secreting the materials that make spicules for the skeleton.

REPRODUCTION AND REGENERATION

Sponges have exceptional powers of regeneration. If a sponge is sieved through a fine silk mesh and the cells left undisturbed, the cells will come together to reform a complete sponge. Porifera reproduce asexually by *budding*. A new individual grows from cells on the parent body, breaks off, and grows into an adult sponge. Sponges also reproduce sexually by means of nonmotile egg cells and motile flagellated sperm. Sponges are *hermaphrodites* because one sponge organism produces both egg and sperm. The gametes may be carried out to sea by flowing water currents or fertilization may take place in the central cavity. The zygote develops into an *amphiblastula* or early embryo. The embryo escapes from the cavity through a pore, swims about for a short while, and then attaches itself, settling down to grow into an adult sponge.

Phylum Coelenterata—Jellyfish, Sea Anemones and Their Relatives_____

The next group on the evolutionary scale is the phylum Coelenterata. Like the sponges, the coelenterates are aquatic animals. Species such as jellyfish, Portuguese-man-of-war, sea anemones and corals live in ocean waters; *Hydra* is a freshwater genus. Unlike the sponges, cells that compose the coelenterate body show more specialization and are grouped together into simple tissues.

GENERAL CHARACTERISTICS

In general, the coelenterate body is shaped like a hollow sac composed of two tissue layers. The *ectoderm* covers the outer surface of the body; *Endoderm* lines inner body surfaces. Between these two tissues is a gelatinous mass of undifferentiated material known as the *mesoglea.*

A unique feature of the coelenterates is the *medusa* and *polyp* forms shown by representative classes in this phylum. Fig. 8.5 illustrates the structure of a jellyfish. Note that its body shows radial symmetry. Notice also that the outer surface of the body curves outward very much like a bowl or a bell. The under surface curves inward and is best described as being concave. This type of body shape is known as a *medusa*. Some coelenterates have the medusa shape all of their lives. Other species exhibit the medusa in part of the life cycle. Fig. 8.6 shows a simplified diagram of the medusa. Notice the *coelenteron,* or gastrovascular cavity. Extending downward from the coelenteron is the centrally located *manubrium* where the mouth is positioned. Tentacles hang from the outer edges of the jellylike bowl. The gonads are suspended under the radial canals and open inward in some species and outward in others.

A number of coelenterate species have a body shape resembling a vase or cylinder. This type of body is known as a *polyp*. For some species the polyp is a stage in the life cycle; in others, the polyp is the adult body

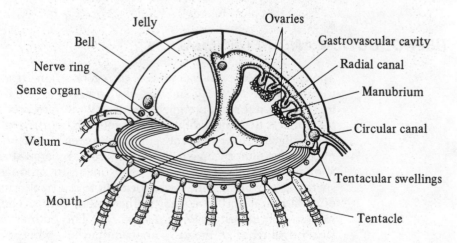

Fig. 8.5 The structure of a jellyfish

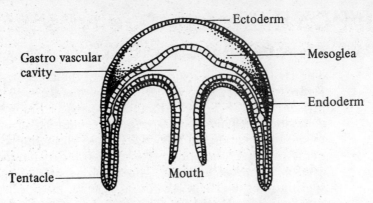

Fig. 8.6 Diagram of a medusa

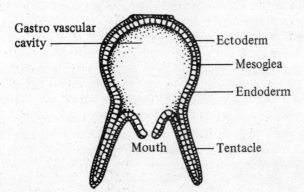

Fig. 8.7 A diagram of a polyp

form. Fig. 8.7 provides a simple diagram of a polyp. Notice that the proximal end is attached. The mouth, surrounded by tentacles, is at the distal end of the animal.

The coelenteron is a distinctive feature of the animals in this phylum. It is a cavity where digestion takes place. Enzymes are secreted into the coelenteron where food is partially digested extracellularly (outside of cells). The partially digested food molecules are then engulfed by cells lining the cavity; these cells complete the digestion intracellularly (inside of cells). In an evolutionary sense, the coelenterate method of digestion is a signpost pointing toward increased specialization of body tissues and organs for digestion. Particles not digested are expelled through the single *mouth-anus*.

Some biologists prefer to use the name Cnidaris for this phylum. This name is derived from the *cnidoblasts* or stinging cells that are in the tentacles of coelenterates. These animals feed on live prey. They capture smaller animals by immobilizing them with toxins secreted by the stinging cells. Each cnidoblast contains a thread capsule called a *nematocyst* which discharges a thread in response to touch or chemical stimulation. The thread may be barbed or coated with toxin. The nematocyst thread either hooks the prey, ensnares it, paralyzes it with toxin, or does all three of these. Once immobilized, the small animal is swept by the tentacles into the mouth-anus.

REPRESENTATIVE CLASSES

Class Hydrozoa

The Hydrozoa are the most primitive of the coelenterates. Many of the Hydrozoa species are colonial, with the members of a colony showing a division of labor. Some of the hydrozoans go through an alternation of generations in which an asexual polyp generation alternates with a sexual medusa generation. In this coelenterate class, the polyp form is dominant. However, the polyps of some species reproduce medusae by the asexual means of budding. The medusae produce gametes (eggs and sperm) and thus begin the sexual generation. Species in the class Hydrozoa include *Hydra, Obelia, Gonionemus* and the Portuguese man-of-war.

Hydra is a freshwater coelenterate that is representative of a genus of the same name. *Hydra* is a polyp and has no medusa form in its life history. In length, this hydrozoan is about 12 millimeters and has about eight tentacles that surround the mouth-anus (Fig. 8.8). Hydras move about by somersaulting, end over end. The animal's locomotion is made possible by cells that have contractile fibers called *myonemes*. These epitheliomuscular cells have locomotor and sensory functions. Reproduction in hydra is sexual and asexual. A single organism produces both egg and sperm which are discharged into the water where fertilization takes place. Asexual reproduction occurs by budding.

Nervous response in *Hydra* is controlled by a simple nervous system. Slim, pointed *sensory cells* scattered throughout the endoderm and ectoderm layers receive stimuli. From the sensory cells, the sensory impulses are passed to nerve cells which form a *nerve net* spread throughout the ectoderm. The nerve net coordinates *Hydra's* activities, enabling the animal to respond to chemical and tactile stimuli in the environment. The nerve net is a very primitive nervous system in which nervous impulses travel in either direction.

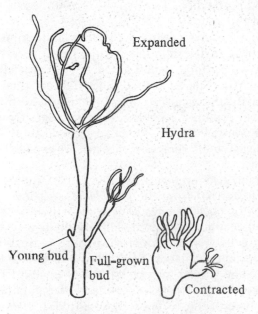

Fig. 8.8   Hydra expanded and contracted

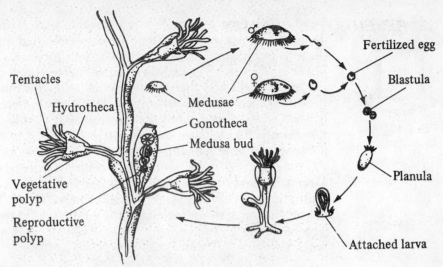

Fig. 8.9 Life cycle of obelia

Obelia is a marine hydroid—a colony of subindividuals. The size of the colony may range from 2.5 to 10 centimeters in height. It is sessile, attached to substrate rocks along the shoreline. Look at Fig. 8.9. You can see that a number of polyps are attached to a stalk. These polyps are specialized for carrying out specific life functions. Some polyps are vegetative and attend to the feeding needs of the group; others have a reproductive function. The *hydrotheca* is a transparent covering that surrounds the vegetative polyps; the *gonotheca* covers the reproductive polyps.

The reproductive polyps do not have tentacles and consequently cannot feed. They produce new polyp individuals by budding. The reproductive polyps produce another kind of bud that resembles a saucer. This bud escapes the polyp through an opening in the gonotheca, becomes free swimming and develops into a medusa, resembling a jellyfish. The medusa is the sexual stage producing egg and sperm. Fertilization takes place in the water. The zygote develops into a ball of cells which changes into a ciliated larva and then becomes a young polyp. The polyp attaches itself and settles down. On maturity, the polyp begins colony building by budding and the cycle repeats. Thus, there is an alternation of an asexual generation with a sexual generation.

Class Scyphozoa—Jellyfish

The Scyphozoa represent the true jellyfish. These coelenterates are free-swimming and inhabit marine waters. In size, a jellyfish may be as small as 2 centimeters in diameter or may be an emormous 4 meters across with trailing tentacles about 10 meters in length. As shown in Fig. 8.5, a jellyfish is the medusa body form with notches in the margin of the bell. Four *oral arms* extend from the mouth opening. Surrounding these arms are four *gastric pouches* where digestion occurs. A complex of canals radiate through the medusa. Attached to the membranes of the canals and encircling the gastric pouches are the gonads.

In some species of jellyfish a polyp stage occurs, but it is subordinate and inconspicuous compared to the medusa. Reproduction by the polyp is asexual, accomplished by a kind of terminal budding known as *strobilation*. Reproduction by the medusa is sexual.

Fig. 8.10 Sea anenome

Jellyfish show greater cell specialization than *Hydra* or *Obelia*. Underlying the ectoderm are true muscle cells that propel the animal through water by regular contractions. Sensory nerves connect with the fibers of the nerve net and serve as the nerve supply for the tentacles, the muscle cells and the sense receptors. Now for the first time we see the emergence of true sense organs: *statocysts* and *ocelli*. The statocyst is a sense organ specialized for receiving and coordinating information that enables an organism to orient itself in respect to gravity. Statocysts are spaced around the margin of the bell. Each statocyst is composed of a circle of *hair cells* which surround a central hardened crystal of calcium carbonate known as the *statolith*. In response to movement of the statolith, the hair cells send impulses to the nerve fibers. The organism can adjust to an up or down position in the water as indicated by the statocyst. The ocelli are groups of light receptor cells located at the base of the tentacles.

Class Anthozoa—Sea Anemones

This class includes the sea anemones, sometimes called "animal-flowers," because of their brightly colored tentacles (Fig. 8.10), and the coral-building animals. Both of these groups are polyps and have no medusa stage. They are sessile, attached to substrate rocks at the edge of the sea. Anthozoans have numerous tentacles that surround an elongated mouth which opens into a tube (stomodaeum) that extends into the gastrovascular cavity. Reproduction may be asexual by budding or sexual involving eggs and sperm.

The epithelial cells of the coral-building coelenterates secrete calcium carbonate walls in which the living polyps hide themselves. The compounds secreted by these polyps build limestone coral reefs.

Phylum Platyhelminthes—Flatworms

The tissues and the organs of the flatworm are developed from the three primary germ layers: ectoderm, mesoderm and endoderm. The simplest of the flatworms demonstrate bilateral symmetry. This phylum represents a step up the evolutionary scale showing definite development of excretory, nervous and reproductive systems. Most of the flatworms are hermaphrodites, and most are parasites, several serious parasites of humans and other animals. Among the flatworms are the planaria, flukes, and tapeworms.

REPRESENTATIVE CLASSES

Class Turbellaria—Planaria

Planaria is a genus of small, freshwater, free-living flatworms that are studied extensively in biology classes. Fig. 8.11 illustrates the general external structure of a planarian. The digestive system consists of a ventral mouth-anus, a pharynx that can be pushed outward, a digestive cavity and an intestine. A true body cavity (coelom) is not present in *Planaria*

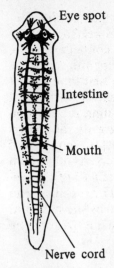

Fig. 8.11
Planaria

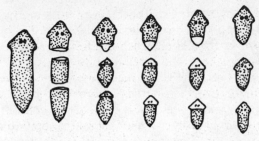

Fig. 8.12   Regeneration in planaria

as indicated by lack of an anus separate from the mouth. The excretory system of *Planaria* consists of a network of branching tubes. The outer ends of these tubes open to the outside through an excretory pore. The inner portion of the tube connects to ciliated cells called *flame cells*, which remove excess water from spaces around the cells.

A consequence of bilateral symmetry is the refinement of body systems. The nervous system of *Planaria* represents an evolutionary advancement. The nerve net of *Hydra* is replaced by two lateral nerve cords extending longitudinally from the anterior end to the posterior end of the body. The presence of anterior *ganglia*, or groups of nerve cells, to sort and coordinate nerve impulses is a signpost to *cephalization*, the development of the head region. The eyespots or *ocelli* can distinguish light from dark and can also discern the direction from which the light comes. The head region has many *chemoreceptors* which aid in the locating of food.

Of special interest to students of biology is the regenerative powers of *Planaria*. Fig. 8.12 illustrates the regeneration story.

Class Cestoda—Tapeworm

Taenia solium is the species name of the tapeworm that infects pigs and people. Having no mouth or digestive system, the adult worm obtains nourishment by absorption through its body wall from the human intestine where it lives. The body of *Taenia* consists of a head, called the *scolex*, attached to a neck which is followed by a series of body segments called *proglottids*. The scolex, which is about 2 millimeters in diameter, is fitted with hooks and four suckers which enable the worm to attach itself to the intestine wall. The proglottids bud from the neck and become progressively more mature and larger as the chain of segments moves toward the posterior end of the animal. A chain of proglottids may be two to three meters long. Each proglottid has a complete reproductive system producing both egg and sperm. The nervous and excretory systems extend through the chain of proglottids (Fig. 8.13).

The life cycle of the tapeworm involves two hosts. A mature proglottid contains a sac filled with hundreds of fertilized eggs. When the proglottid walls rupture, the ground becomes infected with fertilized eggs. If these eggs are ingested by a pig, the protective walls surrounding each egg are digested, releasing developing embryos of the tapeworm into the digestive system of the pig. These embryos bore into the pig's capillaries and are carried by the blood to the muscles where the scolex forms a cyst. The worm now remains encysted in the muscles (meat) of the pig. If the butchered pig (now called pork) is improperly cooked and eaten by a human, the encysted worm becomes activated. Its head begins to bud proglottids and the cycle of infection repeats.

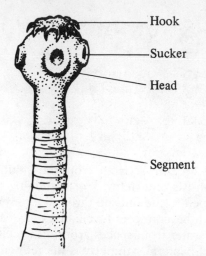

Fig. 8.13 Tapeworm

THE PARASITIC WAY OF LIFE

By definition, a parasite is an organism that lives on or inside of the body of a plant or animal of another species and does harm to the host. Parasites offer physical discomfort to the host and tend to kill slowly, meanwhile having had time to reproduce themselves for several generations. *Ectoparasites* live on the host's body: body lice, dog fleas, ticks. *Endoparasites* live within the hosts's body and exhibit several adaptations for life in an intestine or in muscle or in the blood.

Phylum Nematoda—Roundworms

The roundworms, also called thread worms, are widely distributed, living in the mud of salt and fresh water and in soil. Roundworms are small, tapered at both ends; the elongated body is covered with an enzyme-resistant cuticle. The free-living species have well-developed sense organs, including eyespots and complex mouthparts; parasitic forms are simpler in structure. Anterior ganglia connect with dorsal and ventral nerve cords, articulating with smaller branching nerve fibers. Male and female gametes are produced by separate sexes.

Ascaris lumbricoides is a roundworm parasite that lives in the human intestinal tract. It ranges in length from 15 to 40 centimeters. The eggs escape through the feces of an infected individual and contaminate soil or water. People become infected with *Ascaris* by eating contaminated food.

Hookworms are small nematodes, measuring about seven centimeters in length. Well-developed hooks surround the male genitalia and are present in the mouths of both sexes. Embryos in contaminated soil bore through the skin of the feet and travel through the blood vessels to the lungs. Then they bore through the lung tissue into the bronchi and the windpipe into the throat. They are swallowed and pass into the intestines where they become attached to the intestinal wall. Hookworms drain blood from the host and often cause a severe anemia.

Phylum Annelida—Segmented Worms

An important indicator of evolutionary advancement is the development of the body cavity known in technical language as the *coelom*. Species in the phylum Annelida possess a true coelom. The coelom is a body cavity that has two openings, beginning with an anterior *mouth* and terminating in a posterior *anus*. It is lined with endoderm and separates the internal body organ systems from the muscles of the body wall. The coelom allows space for the development of complex and specialized systems: circulatory, digestive, reproductive and excretory.

GENERAL CHARACTERISTICS

The annelids are segmented worms that live in soil, fresh water, or the sea. Most of the annelids are free-living, although some of the marine forms burrow in tubes and some species (class Myzostoma) are parasites on echinoderms.

The body of an annelid is divided into a series of similar segments and is said to be *metamerically segmented*. Most annelids have a closed *circulatory system* where the blood is contained in vessels. Enlarged muscular blood vessels function as hearts and pump the blood through the system of vessels. Annelids may be *dioecious* (have separate sexes) or hermaphroditic. Most annelid species go through a ciliated larval stage known as the *trochophore* larva. This is a larva of evolutionary importance because the same type appears in several phyla.

REPRESENTATIVE CLASSES

Class Polychaeta—Sandworms

Nereis inhabits burrows in sand and rocks at the edge of the sea (Fig. 8.14). The body of the worm is markedly segmented. The head end is well-developed. It has a muscular pharynx that can be extended outward to capture food. The pharynx is equipped with a pair of hard curved jaws which are designed for grasping. The head bears four simple eyes, four short tentacles, and two longer ones. Each body segment following the head has fleshy protruding appendages called *parapodia* from which grow bristles, or *setae* (*seta*, sing.).

Fig. 8.14   Nereis, the sandworm

The digestive system of *Nereis* is a straight tube consisting of a pharynx, esophagus and a stomach-intestine, where the major part of digestion takes place. Undigested food is eliminated through the anus. Oxygen is taken into the body through the skin covering the parapodia, and it diffuses through the thin walls of the blood vessels. The blood of *Nereis* contains the red pigment hemoglobin and is enclosed in muscular vessels. A pair of coiled *nephridia* in each segment filter out waste materials. Well developed ganglia in the dorsal part of the head relay nerve signals from the sense organs to the ventral nerve cord. Lateral branching nerves from the ventral nerve innervate the body organs. The sexes are separate. Mature gametes are discharged into the water where fertilization takes place. The zygote develops into a trochophore larva.

Class Oligochaeta—Earthworm

Lumbricus terrestis is representative of the earthworms and is studied extensively in biology classrooms. The earthworm burrows in moist soil, feeding on organic materials in the earth (Fig. 8.15). The body is segmented and may have over one hundred metameres. Its under-developed head without eyes and tentacles is well suited to a burrowing way of life.

The digestive system consists of a mouth, a pharynx, a long narrow esophagus, a thin-walled crop, a muscular gizzard, an intestine and an anus. The intestine has a dorsal infolding called the *typhlosole*, an adaptation for increasing surface area necessary for absorptive purposes. The respiratory system of *Lumbricus* is similar to that of *Nereis* except in the earthworm oxygen diffuses through moist body skin and not through parapodia. Paired nephridia are present in each body segment and are used to filter out wastes from the coelomic fluid. Excretion of these wastes takes place through the *nephrostome*, a ventral excretory pore. In the skin, there are sense organs that function as light and touch receptors. The ventral nerve cord with its branching nerve fibers and the anterior ganglia are not unlike that of the sandworm.

Lumbricus is a hermaphrodite. However, earthworms do not self-fertilize; they exchange sperm during copulation. The *clitellum*, a smooth circle of tissue on the outside of the body, is the place where copulating worms attach to each other.

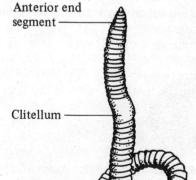

Anterior end segment

Clitellum

Posterior end

Fig. 8.15 Earthworm

Phylum Mollusca—Clams and Their Relatives

The mollusks include the chitons, snails, clams, scallops, squids and octapuses. This is one of the largest animal phyla and includes about 1,000 species.

GENERAL CHARACTERISTICS

Mollusks are generally soft-bodied, nonsegmented, and usually enclosed within a calcium carbonate shell. They are most abundant in marine waters, although some species inhabit fresh water and others live on land. All mollusks have a *mantle*, a flattened piece of tissue which covers the body and which may secrete the calcareous shell. The body of the mollusk is described as being a *head-foot*, a muscular mass having different shapes and functions in the various classes. Between the body and the mantle is the *mantle cavity*, which functions in respiration. A large *visceral mass* contains most of the body organs. Water enters through an *incurrent siphon* and is expelled through an *excurrent siphon*.

The mollusks are the most highly developed of the nonsegmented animals and are considered to be the most advanced invertebrates. They have well-developed digestive, respiratory, excretory and reproductive systems. The circulatory system includes a two-chambered heart equipped with an auricle and a ventricle. Oxygen-laden blood is pumped both anteriorly and posteriorly through two arteries to parts of the body. Blood carrying respiratory wastes is collected through a vein that is closely applied to the nephridia. The deoxygenated blood is transported back to the gills. The excretory system has a pair of nephridia that excrete filtered wastes through an excretory pore into the mantle cavity. The excurrent siphon removes the dissolved wastes out of the shell.

The nervous system contains three pairs of ganglia positioned near the esophagus, in the foot, and at the end of the visceral mass. These ganglia are connected by transverse nerve fibers. Simple sense receptors in the form of sensory cells sensitive to light and touch are positioned in the margin of the mantle.

In most mollusks, the sexes are separate. The reproductive system is a mass of gland tissue that lies in the muscular foot near the coiled intestine. In some species sperm are conducted through the excurrent siphon of the male into the mantle cavity of the female by way of the incurrent siphon, and fertilization takes place in the mantle cavity of the female. In other species, eggs and sperm are released into the water where fertilization takes place. Some mollusks—oysters, scallops, and primitive worm-like forms—are hermaphroditic.

The fertilized eggs of clams and other bivalves develop into a larval stage known as a *glochidium*. The glochidium is discharged into the water and attaches as a parasite to the gills of fish, where it stays until reaching maturity.

Table 8.1 summarizes the important characteristics of the major classes of mollusks. One class—that of the bivalves—is discussed in more detail.

TABLE 8.1.   Phylum Mollusca

Class	Examples	Characteristics
Amphineura	chiton	Marine; bilaterally symmetrical; ventral foot; shell of eight calcareous plates; gills in mantle cavity.
Pelecypoda (Bivalvia)	clam, mussel, oyster, scallop, shipworm	Marine and freshwater; body enclosed in right and left shells; head reduced; filter feeders; compressed foot; sexes separate or hermaphroditic.
Gastropods	snail, whelk, limpet, slug	Marine, freshwater, land; most with spiral, single shells; well-developed head having tentacles and eyes; foot for locomotion; shell absent in slugs; trochophore larva.
Cephalopods	squid, octopus nautilus	Marine; well-developed head with large eyes; shell absent or present; tentacles; siphon used for locomotion; ink gland; sexes separate.

A REPRESENTATIVE CLASS—PELECYPODA (BIVALVIA)

The clam is an excellent representative of this class, usually known as the bivalves because of the presence of two valves or shells. The right and left valves are hinged on the dorsal side and held tightly together by muscles attached to the inner surfaces of the shells. Lining these inner surfaces is the membranous mantle. The cavity inside of the shells is the mantle cavity.

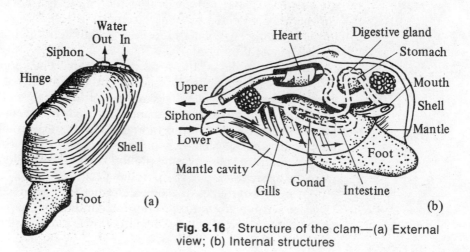

Fig. 8.16 Structure of the clam—(a) External view; (b) Internal structures

The clam has a large muscular foot which can be thrust between the opening of the shells and which contracts and expands during locomotion due to the action of protractor and retractor muscles attached to the inner faces of the shells. There is a large visceral mass at the base of the foot containing most of the animal's organs.

Four parallel gill plates are adjacent to the visceral mass. These gills function in respiration. Oxygen from the water in the mantle cavity diffuses into the gills, which are surrounded by capillaries. Oxygen diffuses into the capillaries and then into the bloodstream. Carbon dioxide diffuses out into the surrounding water by way of the capillaries and the gills.

The digestive system of the clam is complete. A mouth occupies the anterior end of the visceral mass. Leading from the mouth is a short esophagus followed by a stomach and a coiled intestine extending partly into the foot. At the end of the intestine is an anus. A bilobed digestive gland is on either side of the stomach (Fig. 8.16).

Phylum Echinodermata—Starfish and Their Relatives

Echinoderms are spiny-skinned invertebrates that include the starfish, brittle stars, sand dollars, sea urchins and sea cucumbers. Although they do not look very much like vertebrate animals, the development of the echinoderm embryo strongly resembles that of the chordates in the early stages. The larval stage is free-swimming and shows bilateral symmetry.

GENERAL CHARACTERISTICS

All echinoderms live in the sea. Most species are capable of a very slow, creeping locomotion. The only group of sessile echinoderms is the sea lilies.

The name echinoderm means spiny-skin, a distinctive feature of the phylum members. Just under the skin, calcareous spines and plates form a skeleton. Another distinctive characteristic of the echinoderms is pentaradial symmetry: the body is built on a plan of five *antimeres* radiating from a central disc in which the mouth is in the middle. The digestive system is complete, although the anus does not function. The echinoderms have no head and no excretory and respiratory systems. They do, however, have a *water vascular system* composed of a series of fluid-filled tubes that are used in locomotion. Changes of pressure in this system enable an echinoderm to extend and retract *tube feet*. The tube feet are used in locomotion and in some species they are used to capture prey. In the echinoderms, the sexes are separate.

Table 8.2 summarizes the important characteristics of the classes of echinoderms. One class—that of the starfish—is discussed in more detail.

TABLE 8.2. Echinoderms

Class	Examples	Characteristics
Crinoidea	sea lily, feather	Sessile, attached by a stalk; branched arms; ciliated tube feet used for feeding; some species are free swimming; more abundant during Paleozoic era.
Asteroidea	starfish	Free moving by means of tube feet; arms branching from a central disc.
Ophiuroidea	brittle stars, serpent stars, basket stars	Free moving; thin flexible arms marked off from disc; tube feet used as sensory organs and for feeding.
Echinoidea	sand dollar, sea biscuit, sea urchin	Free moving; body fused plates or flattened disc, without free rays, covered with calcareous plates; some species covered with spines.
Holothurioidea	sea cucumber	Free moving; elongated flexible body with mouth at one end; sometimes with tentacles; skeletal elements of the skin reduced.

A REPRESENTATIVE CLASS—ASTEROIDEA (STARFISH)

The starfish is an excellent representative of the echinoderms. It has all of the distinguishing characteristics: pentaradial symmetry, spiny skin, tube feet controlled by a water vascular system, no head, excretory or respiratory system. Protruding from the wall of the coelom and extending out between the calcareous plates into the sea water, are the *papulae*, sac-like structures that function as respiratory and excretory organs.

The mouth is located in the center of the disc on the underside of the body. The mouth side of the body is called the oral side. The aboral side is the upper surface without the mouth. A short esophagus leads from the mouth to the cardiac portion of the stomach. A constriction in the stomach wall separates the cardiac portion of the stomach from the pyloric part. The cardiac stomach is turned inside out and pushed through the mouth when the starfish is eating. The stomach engulfs the food, usually molluscs or crustaceans, and digests it before pulling the stomach back to the inside. The intestine and the anus of the starfish are practically nonfunctional.

On the aboral side of the starfish is a colored plate known as the *madreporite*. Water enters the starfish through minute openings in this plate. Water is drawn by ciliary action down into the stone canal (made rigid by calcareous rings) to the ring canal that encircles the central disc. The ring canal has five radiating canals which extend into the arms of the starfish. Short side branches connect the radial canals with many

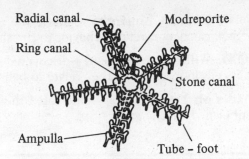

Fig. 8.17 Water vascular system of the starfish

pairs of tube feet which contract and expand in response to the water pressure in the *ampulla*, a muscular sac at the upper end of the tube feet (Fig. 8.17).

The nervous system is composed of a nerve ring located in the disc from which a ventral and radial nerve branch into each arm. The radial nerves have finer branches which extend throughout the body. At the tip of each arm is a light sensitive eyespot that is innervated by the radial nerve.

Starfish sexes are separate. Paired gonads are to be found in each ray. The eggs of the female and sperm of the male escape through pores on the aboral surface of the starfish. Fertilization takes place in the water. During embryonic development, starfish pass through several larval stages.

Starfish have remarkable powers of regeneration. If an arm breaks off, the arm grows back. Should a piece of the central disc be attached to the amputated arm, a new individual will grow from the dismembered part. Starfish prey on oysters. At one time, oyster "farmers" would clear the oyster beds of starfish, cut them up and throw the cut pieces back into the water. What they accomplished was the increase in the numbers of starfish. They, in effect, aided the process of starfish regeneration.

Chronology of Famous Names in Biology

1907   **H. V. Wilson** (American)—discovered that individual cells of the sponge, if left undisturbed in a culture dish, will reform sponge aggregates.

1929   **Charles M. Yonge** (Scots)—wrote a treatise on the biology of coral reefs in which he elucidated the patterns of behavior of coral reef invertebrates.

1939   **Ernest E. Just** (American)—developed basic research techniques to study eggs of marine animals.

1940   **Libbie Hyman** (American)—noted authority on the invertebrates, demonstrated metabolic gradients in *Planaria*.

1965   **T. L. Lentz** and **R. J. Barrnett** (American)—elucidated the fine structure of the *Hydra* nervous system; demonstrated how the feeding response is stimulated in *Hydra*.

1967   **Donald Kennedy** (American)—found the effects of the abdominal ganglia on reflex behavior in the slug.

1971   **A.O.D. Williams** (American)—discovered the function of brain cells in mollusks as they relate to behavior.

1975   **Charles M. Yonge** (Scots)—elucidated the life cycle of the giant clam *Tridocna gigas*.

Words for Study

amphiblastula
ampulla
anterior
antimeres
blastula
cephalization
chemoreceptors
clitellum
cnidoblast
coelenteron
coelom
collar cells
dioecious
distal
dorsal
ectoderm
ectoparasite
endoderm
endoparasite
filter feeder
flame cells

ganglia
glochidium
gonotheca
hair cells
hermaphrodite
hydrotheca
lateral
madreporite
mantle
manubrium
medusa
mesenchyme cells
mesoderm
mesoglea
metameres
myoneme
nematocyst
nephridia
nephrostome
ocellus
papulae

parapodia
parthenogenesis
polyp
posterior
primary germ layer
proglottids
proximal
regeneration
scolex
setae
spicules
spongin
statocyst
statolith
strobilation
symmetry
trochophore larva
typhlosole
ventral

Questions for Review

PART A.   Completion. Write in the word that correctly completes each statement.

1. Relative positions of body organs on opposite sides of a dividing line are described in terms of ..1..

2. The portion of the fish fin that is attached to the body is the ..2.. end.

3. Invertebrates are animals without ..3..

4. Budding is a form of ..4.. reproduction.

5. The name Porifera means ..5.. bearing.

6. Needle-like crystals that make up the skeleton of a sponge are called ..6..

7. An organism that produces both egg and sperm is known as an ..7..

8. Cnidoblasts are ..8.. cells.

9. A thread capsule found in coelenterate tentacles is the ..9..

10. *Hydra* lives in ..10.. water.

11. The polyps in a colony of *Obelia* reproduce asexually by ..11..

12. Orientation to gravity is controlled by jellyfish sense organs known as ..12..

13. *Taenia* is the genus of the ..13..

14. *Ascaris* is a parasite belonging to phylum ..14..

15. Infection by hookworm can be avoided by wearing ..15..

16. The true coelom first appears in the ..16..

17. Animals in which the sexes are separate are said to be ..17..

18. The body of all mollusks is covered by a membranous ..18..

19. The incurrent and excurrent siphons are characteristic of species in the phylum ..19..

20. Tube feet are characteristic of the phylum ..20..

PART B.   Multiple Choice. Circle the letter of the item that correctly completes each statement.

1. Bilateral symmetry is characteristic of animals that have a
   - (a) shell and plates
   - (b) head and tail
   - (c) foot and coelom
   - (d) spines and tube feet

2. Distal is a body directional term that refers to
   - (a) a forward end
   - (b) a backward end
   - (c) an attached end
   - (d) a free end

3. In animal cells, excess nutrients are stored as
   - (a) glucose and glycogen
   - (b) glucagon and fatty acid
   - (c) glycogen and fat
   - (d) glycogen and protein

4. The growing back of a lost part is known as
   - (a) regeneration
   - (b) budding
   - (c) vegetative propagation
   - (d) parthenogenesis

5. A true statement about sponges is
   - (a) the cells are organized into tissues
   - (b) they move quite rapidly
   - (c) the digestive system is well developed
   - (d) the cells show a division of labor

6. Types of sponge cells that can change their form and function are
   - (a) collar cells
   - (b) flame cells
   - (c) mesenchyme cells
   - (d) epithelial cells

7. The coelenterate body is composed of
   (a) only undifferentiated cells
   (b) one tissue layer
   (c) two tissue layers
   (d) three tissue layers

8. The coelenteron is a
   (a) foot
   (b) tentacle
   (c) sac
   (d) cavity

9. Medusae produce by
   (a) budding
   (b) eggs and sperm
   (c) regeneration
   (d) parthenogenesis

10. Alternation of generations is demonstrated in the life cycle of
    (a) *Hydra*
    (b) sea lily
    (c) pelecypods
    (d) *Obelia*

11. Coral reefs are formed by
    (a) lime-secreting polyps
    (b) the bodies of sea anemomes
    (c) the shells of foraminifera
    (d) jellyfish statocysts

12. Flatworm embryos differentiate into
    (a) ectoderm, mesoglea, mesoderm
    (b) ectoderm, mesoglea, endoderm
    (c) ectoderm, mesoderm, endoderm
    (d) ectoderm, mesenchyme, endoderm

13. Bilateral symmetry brings with it
    (a) refinement of body systems
    (b) the development of the nerve net
    (c) increased reproduction
    (d) alternation of generations

14. Each tapeworm segment has a complete
    (a) digestive system
    (b) nervous system
    (c) reproductive system
    (d) locomotor organ

15. A true statement about *Planaria* is
    (a) *Planaria* are parasites.
    (b) *Planaria* lack powers of regeneration.
    (c) *Planaria* show the first signs of cephalization.
    (d) *Planaria* have segmented bodies.

16. The circulatory system of the annelid is
    (a) open
    (b) closed
    (c) partly closed
    (d) varying

17. Metamerism refers to
    (a) male and female gonads in the same animal
    (b) a parasitic infestation of humans
    (c) an evolutionary trend in roundworms
    (d) a series of identical segments

18. The trochophore larva
    (a) is a fossil remnant
    (b) shows relationships between phyla
    (c) is part of the roundworm life cycle
    (d) has free flowing pseudopods

19. A true statement about the earthworm is
    (a) Earthworms are self-fertilized.
    (b) The earthworm sexes are separate.
    (c) Earthworms copulate and exchange sperm.
    (d) Earthworms copulate and exchange eggs.

20. The papulae in the starfish function as organs of
    (a) locomotion   (c) respiration
    (b) reproduction   (d) digestion

PART C.   Modified True-False. If a statement is true, write "true" for your answer. If a statement is incorrect, change the <u>under-lined</u> word to one that will make the statement true.

1. A true body cavity is lined with <u>mesoderm</u>.

2. The dorsal region of the body refers to the <u>under</u> surface.

3. Contractile proteins are associated with <u>feeding</u>.

4. *Coelenterates* are the first group to show <u>tissue</u> organization.

5. Sponges have an <u>external</u> skeleton.

6. Spongin is a fibrous <u>carbohydrate</u>.

7. The bowl shape of a jellyfish is known as an <u>umbrella</u>.

8. The body form of hydra is a <u>polyp</u>.

9. The nerve net is associated with <u>starfish</u>.

10. Hair cells are part of the <u>ocelli</u>.

11. "Animal-flowers" refer to <u>jellyfish</u>.

12. The function of flame cells is related to <u>digestion</u>.

13. Proglottids are segments in <u>Planaria</u>.

14. The head of a tapeworm is known as the <u>proglottid</u>.

15. *Taenia* lives in two hosts: human and the <u>snail</u>.

16. Annelids are <u>round</u> worms.

17. *Nereis* is the genus name of the <u>tapeworm</u>.

18. The sandworm has a well developed <u>head</u>.

19. The most advanced phylum of invertebrates is the <u>annelids</u>.

20. The visceral mass is a characteristic of species belonging to the phylum <u>Nematoda</u>.

Answers to Questions for Review

PART A

1. symmetry
2. proximal
3. backbones
4. asexual
5. pore
6. spicules
7. hermaphrodite
8. stinging
9. nematocyst
10. fresh
11. budding
12. statocysts
13. tapeworm
14. Nematoda
15. shoes
16. flatworms
17. dioecious
18. mantle
19. Mollusca
20. Echinodermata

PART B

1. b
2. d
3. c
4. a
5. d
6. c
7. c
8. d
9. b
10. d
11. a
12. c
13. a
14. c
15. c
16. b
17. d
18. b
19. c
20. c

PART C

1. endoderm
2. back
3. movement
4. true
5. internal
6. protein
7. medusa
8. true
9. *Hydra*
10. statolith
11. sea anenomes
12. excretion
13. tapeworms
14. scolex
15. pig
16. segmented
17. sandworm
18. true
19. mollusks
20. Mollusca

CHAPTER 9

INVERTEBRATES: THE ARTHROPODA

The Arthropoda is the largest animal phylum in numbers of individuals, encompassing approximately 800,000 species, more than all of the other animal species combined. It includes spiders, ticks, mites, lobsters, crabs, insects, centipedes and millipedes (Fig. 9.1). Arthropods are widely distributed, occupying habitats in marine, freshwater and terrestrial environments.

The name *arthropod* in literal translation means "jointed foot," a distinctive characteristic of this group, expressed traditionally as "*jointed appendages*". The arthropods are segmented animals protected by an exoskeleton made of protein and the flexible but tough carbohydrate chitin. The chitinous exoskeleton is fashioned in articulating plates held together by hinges covering both the body and the appendages, and attached to muscles that make possible quick and unencumbered movements.

The body cavity (coelom) is reduced in size and in its place is a *hemocoel* (blood cavity), a part of the *open circulatory system* in which a pulsating section functions as a dorsal heart. *Hemolymph* bathes the body organs directly because it is not confined to vessels. Respiration in many arthropod species makes use of *tracheal tubes* which communicate to the external environment by *spiracles*, pores that appear one pair to a segment. Some arthropods use gills as organs of respiration. Nervous impulses are carried on a ventral nerve cord occupying a mid-position in the body. The sexes are separate. Some of the insects reproduce parthenogenetically, a process in which eggs develop without fertilization.

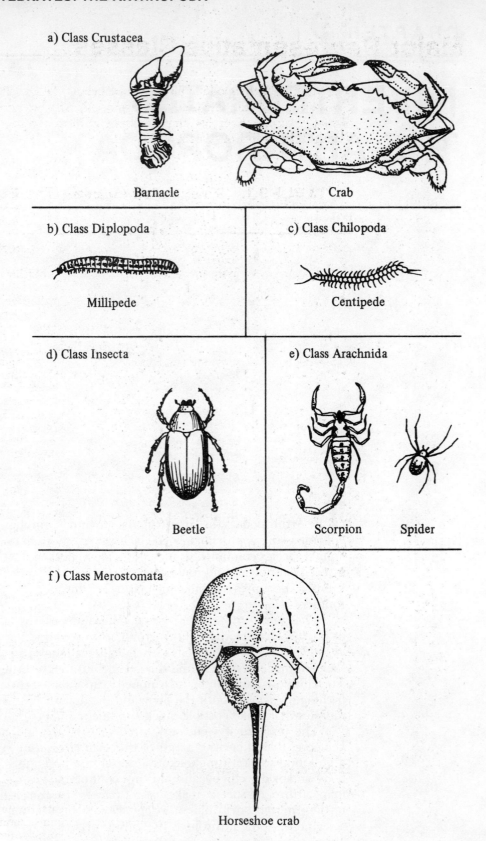

a) Class Crustacea

Barnacle   Crab

b) Class Diplopoda

Millipede

c) Class Chilopoda

Centipede

d) Class Insecta

Beetle

e) Class Arachnida

Scorpion   Spider

f) Class Merostomata

Horseshoe crab

Fig. 9.1 Some representative classes of arthropods

Major Representative Classes_____

The phylum Arthropoda is divided into several classes. The Arachnida, Crustacea, Insecta, Diploda, and Chilopoda are the most important. Table 9.1 summarizes the characteristics of these classes, while a more detailed discussion of certain classes follows.

TABLE 9.1.   Phylum Arthropoda—The Easy Way

Class	Examples	Characteristics
Arachnida	spider, tick, mite, scorpion, horseshoe crab, harvestmen	Mostly terrestrial; small-to-moderate size; body divided into cephalothorax and abdomen; 6 pairs of appendages; 4 pairs of walking legs; no antennae; simple eyes; book lungs or trachae.
Crustacea	crab, shrimp, lobster, barnacle	Mostly aquatic; marine; gill breathers; exoskeleton of chitin; chewing mouthparts; 2 pairs of antennae; 3 or more pairs of legs.
	crayfish, water flea sow bug	freshwater terrestrial
Insecta	grasshopper, dragonfly, cockroach, termite, louse, flea, fly, butterfly, bee, beetle	Mostly terrestrial, freshwater, no marine forms; body divided into head, thorax, abdomen; mouthparts for biting, sucking, lapping; 2 pairs of wings; 3 pairs of legs; breathing by tracheae; excretion by Malpighian tubules; one pair of antennae; 2 compound eyes.
Diplopoda	millipede	Terrestrial; herbivores; head with antennae and chewing mouthparts; body segmented; 2 pairs of walking legs ventral on each segment; 2 eyes; breathing by trachea; 2 pairs of spiracles on each segment.
Chilopoda	centipede	Terrestrial; carnivorous; body segmented; one pair walking legs lateral on each segment; one pair long antennae; poison fangs on first body segment; chewing mouthparts.

CLASS ARACHNIDA—SPIDERS

By careful examination of a spider, you would be able to note the major external characteristics. The body is divided into two regions: the *cephalothorax* and the *abdomen*. The cephalothorax, an arthropod characteristic, is the fusion of the head and the thorax. The cephalothorax of the spider supports six pairs of jointed appendages. The first appendage has been modified into jaws called *chelicerae*. The second are the *palps*, sense receptors and grasping organs, leg-like in females but bulbous in males. The remaining appendages are four pairs of walking legs characteristic of arachnids.

In spiders in the posterior end of the abdomen lying underneath (ventral to) the anus are several pairs of rounded projections. These are called the *spinnerets*. They contain a group of flexible tubules through which silk secreted by the silk glands leaves the body when the spider spins a web.

On the anterior dorsal surface of the cephalothorax are eight simple eyes. These are the external parts of specialized systems that control the physiological processes of the spider. Nerve fibers connect the eight eyes with a nerve mass that surrounds the esophagus. This is where the ventral and dorsal ganglia unite. Branching nerve fibers service the various parts of the body.

The food of the spider consists of body juices of other animals. These juices are drawn through the mouth and the esophagus by action of a sucking stomach, which then passes the food on to another stomach where digestion takes place. Digestion is aided by five pairs of pouched glands. The digestive tract ends in a sac-like rectum and an anus.

Most spiders breathe by means of *book lungs*. These are chambers or cavities containing many thin, hollow membranous plates. The hollow spaces are connected to the outside by fine tracheal tubes. Oxygen is distributed by the blood circulating around the hollow spaces (sinuses). The heart is in the dorsal part of the abdomen fitted into the *pericardial cavity*. Blood enters the heart through a pair of valve-like openings and is pumped out through vessels which empty into the body spaces.

Excretion in the spider is controlled by a pair of *Malpighian tubules*. These are long slender tubules attached at one end to the digestive tract. Nitrogen-containing wastes in the body fluid are changed into uric acid which is then moved through the Malpighian tubule to the end of the digestive tract where it is ultimately excreted as dry crystals (Fig. 9.2).

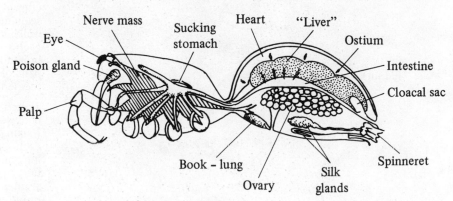

Fig. 9.2 Internal structure of the spider

The sexes in the spider are separate. The sex organs are positioned in the ventral part of the abdomen and communicate with the outside through a ventral opening. During copulation, the male uses modified palps to transfer sperm to the female, effecting internal fertilization. The fertilized eggs are laid in a silk cocoon by some species of spiders. In other species, the eggs are carried around by the female until they are hatched.

CLASS CRUSTACEA—LOBSTERS AND THEIR RELATIVES

Derived from the Latin *cursta* meaning crust, the name Crustacea describes the lobsters and their relatives aptly. The body is covered by a tough exoskeleton arranged in the form of arched plates which thin out at the joints to permit maximum movement. The lobster is representative of this class.

External Characteristics

The segmented body is divided into a cephalothorax and an abdomen. Six segments of the head and eight segments of the thorax are fused into a single cephalothorax; seven segments compose the abdomen. Pairs of jointed appendages are outgrowths of each of the body segments except the first and the last (Fig. 9.3).

The first body segment of the cephalothorax has a pair of *compound eyes* (Fig. 9.4) positioned at the ends of long flexible stalks that can be extended or retracted. The compound eye is an arthropod characteristic that represents an evolutionary development in light receptors. The compound eye is a collection of thousands of light-gathering units, the *ommatidia*, each with its own lens and light-sensitive cells. Each unit makes its own picture of part of an object. These separate pictures are put together in a pattern resembling a collage so that the compound eye perceives thousands of pictures of the same object but at slightly different angles. The compound eye does not render a clear picture of an object but is a very efficient device for spotting movement of prey.

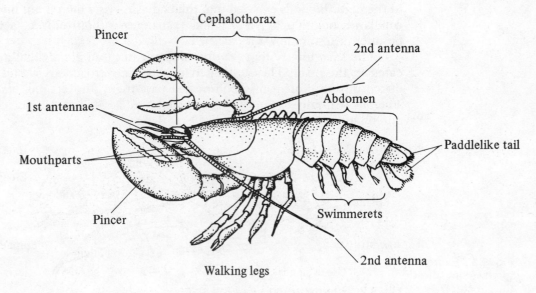

Fig. 9.3 External structure of the lobster

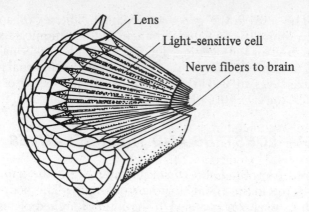

Lens

Light–sensitive cell

Nerve fibers to brain

Fig. 9.4 A diagram of the arthropod compound eye

Segments 2 and 3 on the cephalothorax bear pairs of antennae, sense receptors for touch and chemical stimuli. On segments 4–9 are appendages that have been modified into biting and chewing mouthparts. Segment 10 bears two large pincers (claws) used to fight off enemies and to capture food. Segments 11–14 have four pairs of walking legs. Segments 15–19 are abdominal segments; they bear appendages called *swimmerets* that function mainly in circulating water over the gills. There are two paddle-like appendages on segment 20 which constitute the tail, an appendage modified for swimming.

Internal Structure

The respiratory system consists of feathery *gills*, outgrowths of the body wall that lie on both sides of the body. The gills, surrounded by blood vessels, are bathed by water in the *gill chamber*, a space under the anterior part of the exoskeleton. The gills absorb oxygen from the water and release carbon dioxide into the water.

The heart is located dorsally. Blood enters the heart through three pairs of valve-like openings. It is pumped out through several arteries which branch to all parts of the body, where it delivers oxygen and food to the cells. Blood is emptied into the body sinuses which drain into the capillaries of the gills. Here blood exchanges carbon dioxide, a waste product of respiration, for oxygen.

The excretory system consists of a pair of *green glands* that are located in the head. These glands filter out the nitrogenous wastes produced during protein metabolism and pass them into a small *bladder*. The bladder empties out through a small pore located at the base of the antennae (Fig. 9.5).

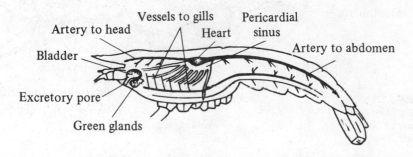

Artery to head

Vessels to gills

Heart

Pericardial sinus

Artery to abdomen

Bladder

Excretory pore

Green glands

Fig. 9.5 Excretory system of the lobster

The digestive system consists of a mouth, a short esophagus, a stomach divided into cardiac and pyloric chambers, an intestine and an anus. The cardiac portion of the stomach contains special grinding organs known as the *gastric mill*. Finely ground food enters the pyloric chamber of the stomach where most of the digestion takes place and where useful nutrients are absorbed into the blood.

The nervous system includes a dorsal "brain" which is really a large ganglion situated in the cephalothorax above the digestive system. These rings of tissue connect with a ventral chain of ganglia from which a double nerve cord extends into the tail. Nerves branch out from the dorsal brain and the ventral cord (Fig. 9.6).

Sexes in the lobster are separate. Bilobed gonads are located ventrally in the cephalothorax. Sperm cells leave the testis by way of *sperm ducts* that open into pores at the base of the fifth pair of walking legs. *Oviducts* in the female conduct eggs from the ovary to openings at the base of the third pair of walking legs. The early embryos develop while they are attached to the swimmerets of the female.

CLASS INSECTA

The Insecta represent the most advanced class of the modern arthropods. Scientists estimate that the numbers of species in this class are probably as high as several million. To date, more than one half million have been described in scientific literature. It is known that there are more kinds of insects than any of the other animals combined. The insects are by far the dominant form of terrestrial life. The diversity of the insects is unmatched.

There are many reasons for the success of the insects. The modern insects are relatively small organisms ranging in size from 1.5 to 50 millimeters. Their small body size and consequently small food requirement has made them quite successful in the struggle for survival. Another reason for their survival is the fact that they have wings and can fly away from predators or toward food sources. (In fact the insects are the only invertebrates that can fly.) That insects have developed adaptations for survival can be attested to by their wide distribution in terrestrial and aquatic environments. Insect species occupy habitats from the equator to the arctic and from sea level to the snow fields of the highest mountains. Many live in fresh water during the larval stages and several species spend their adult lives near the water.

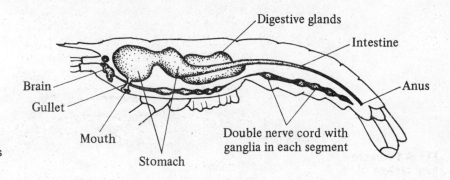

Fig. 9.6 Digestive and nervous system of the lobster

Digestive glands

Intestine

Brain

Anus

Gullet

Double nerve cord with ganglia in each segment

Mouth

Stomach

General Characteristics

There are several characteristics that all insects have in common. The insect body is divided into a well-defined head, thorax and abdomen. Segments 1 to 6 are fused to form the head on which there are a pair of compound eyes and one or more simple eyes. Segment 2 carries a single pair of antennae. The mouthparts vary according to the species of insect. The grasshopper, for example, has biting mouthparts, whereas in the butterfly the mouth is structured for sucking and in the mosquito the mouth serves as a piercing device. Segments 7–9 make up the thorax. Each of these segments bears a pair of walking legs. Most insects bear a pair of wings on the last two segments of the thorax.

The composition and structure of wings differ among orders of insects. For example: the front wings of the beetle serve as protective armor and are generously impregnated with chitin, while the wings of the dragonfly are delicate and membranous. The second wing pair in mosquitoes and grasshoppers is greatly reduced in size and these stubs are used as balancing organs rather than for flight purposes.

The number of segments that make up the insect abdomen varies from a maximum of 11 downward. Although the abdomen bears no appendages, it does have a number of breathing pores (spiracles). The spiracles open into air tubes called trachea that conduct oxygen to all of the body cells.

The life history of an insect may involve several stages in which the young form of the insect does not resemble the parent. Such a life cycle in which there is change of body form is known as *metamorphosis*.

Metamorphosis

Metamorphosis usually involves distinct stages. The life cycle of the butterfly provides an excellent example of four stages: (1) the egg or embryonic stage (2) the *larva* or feeding stage (3) the *pupa* or cocoon stage and (4) the adult stage. The larva is an active stage of life when the feeding organism is an entirely different organism in body form from the parent. The caterpillar, for example, does not resemble the adult butterfly. The pupa is a quiescent nonfeeding stage in the life cycle (Fig. 9.7).

The process of metamorphosis in insects is under control of *hormones* secreted by cells in the brain. Hormones are protein molecules that are secreted by endocrine (ductless) gland cells into the blood for transport to the sites of action. Hormones are specific and stimulate certain *target* cells or organs to grow or to carry out general metabolic activities. *Brain hormone*, secreted by certain cells in the insect's brain, stimulates the *prothoracic gland*, which, in turn, secretes *ecdysone*, a growth and differentiation hormone. Ecdysone controls the molting of the larva and the change into the pupa stage. Another endocrine gland, the *corpus*

Egg   Larva   Pupa   Adult

Fig. 9.7 Complete metamorphosis in the butterfly

allatum, lies near the brain in the larva (caterpillar) and secretes a hormone called *juvenile hormone*. This hormone encourages larval growth and molting (shedding of skin), but prevents the larva from changing body form. After the *corpus allatum* stops secreting juvenile hormone, the metamorphosis to the adult body form occurs.

Representative Orders

The large number of insect species are classified into more than 15 orders. The members of these orders vary in size, feeding habits, habitat, life cycle, and behavior, with some showing complex patterns of social behavior. The grasshopper illustrates the structure of a typical insect, the honeybee the life style of a social insect.

Order Orthoptera—Grasshopper

Fig. 9.8 shows parts of the grasshopper. The head of the grasshopper is clearly defined. There are two lateral compound eyes separated by three simple eyes or ocelli. The single pair of antennae have sensory functions. The grasshopper has biting mouthparts which include a pair of chewing jaws, a pair of flap-like structures for manipulating food, a lower lip-like structure and a tongue-like organ used for the mastication and manipulation of food. The forewings are hard, leathery and opaque. The hind wings are thin, membranous and transparent.

The thorax of the adult grasshopper is divided into three clearly seen parts: the prothorax (with a saddle-like covering called the *pronotum*), mesothorax, and metathorax. The end of the abdomen is modified to form the sex organs. In males, the posterior tip of the abdomen is rounded and the projection used in copulation is present. The posterior end of the female's abdomen is a forked structure, the *ovipositor*, used to dig holes into which the eggs are deposited.

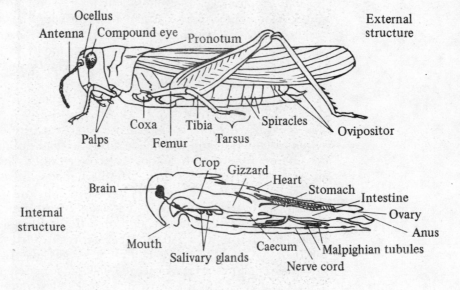

Fig. 9.8 Parts of the grasshopper

The grasshopper has three pairs of legs; one pair each is on the prothorax, mesothorax, and metathorax. The third pair of legs are strong and elongated, adapted for jumping.

The digestive system consists of a mouth that leads into a narrow esophagus which connects with a thick-walled grinding organ, the *gizzard*, which is followed by a thin-walled stomach. The narrow intestine ends with the anus. The cavity of the mouth gets secretions from two salivary glands. Six double-blind sacs surround the gizzard and stomach and empty enzymes into the stomach.

On the first segment of the abdomen, one on each side of the animal, are two *tympanic* membranes. These structures respond to vibrations and sound waves. On two thoracic segments and on eight segments of the abdomen are the spiracles, the external openings of the tracheal tubes and air sacs that branch (ramify) throughout the body. The small branches of the tracheal tubes are in contact with the body cells conducting oxygen to them and removing carbon dioxide wastes from them.

The circulatory system includes a dorsal heart. Blood enters the heart through five pairs of valve-like openings called *ostia* and is pumped into the body sinuses. The hemocoel receives blood from all parts of the body and returns it to the heart. The blood of the grasshopper is merely a circulating medium for food and wastes.

The excretory system consists of numerous fine Malpighian tubules which reabsorb water and excrete protein wastes in the form of dry uric acid crystals.

The sexes in the grasshopper are separate. The male reproductive system consists of two *testes* located near the intestine, a sperm duct, and a copulatory organ situated at the tip of the abdomen. In the female, eggs are formed in two ovaries and travel through an oviduct to the vagina. During copulation, the seminal receptacle of the female receives and stores sperm. When eggs pass through the vagina, they are fertilized by the stored sperm. The female deposits the eggs in the soil by digging a hole with the ovipositor.

The zygote develops quite rapidly into an embryo, but then growth and development stop for a while. This rest period in embryonic growth is called *diapause*, which, in the case of the grasshopper, lasts over the winter. Development resumes in the spring and by early summer young grasshoppers, called *nymphs*, emerge. The nymphs are wingless but, after a series of molts, grow to adult size.

Order Hymenoptera—Honeybee

The honeybee exhibits a specialized social structure called *polymorphism*, a condition in which individuals of the same species are specialized for different functions. In a honeybee colony, three classes of individuals arise: fertile males called *drones*; fertile females, or *queens*; and sterile females, or *workers*. The workers have a special concave surface on the second pair of walking legs called a *pollen basket* used to carry pollen.

The queen bee receives sperm from the drone once during her lifetime. The sperm are stored in a special organ called the *spermatotheca* in which they may live for years. Fertilized eggs give rise to females, most of which remain workers. A special female may be selected by the colony and fed a diet of "royal jelly" which causes her to grow larger than the others and become fertile. This fertile female will become a queen, and either take over the existing colony or start a colony of her own. Drones develop from unfertilized eggs by the process of parthenogenesis.

Importance to Humans

The arthropods have very definite effects on human life. The arachnids influence the quality of human life in diverse ways. Spiders are predators on insects that may offer discomfort or harm to humans, and some feed on decomposed organic matter and serve as decomposers. On the other hand, ticks cause Rocky Mountain spotted fever and mites are spoilers of grain. The crustaceans for the most part have a great deal of value. Crabs, lobsters and shrimp serve the cause of human nutrition deliciously.

Insects cannot be avoided. Some insects are helpful and others are harmful. Honeybees provide honey and pollinate flowers. Ladybird beetles destroy other insects that are crop destroyers. Many insects such as flies, fleas, and mosquitoes are carriers of disease; others such as Japanese beetles, tent caterpillars and grasshoppers cause serious damage to foliage and food crops. A greater-than-billion-dollar industry has been built for exterminating insects that infest our homes and gardens. Insects help in the balance of nature and also in certain situations help to unbalance natural communities. This very successful species is human's greatest competitor for food on the earth.

Chronology of Famous Names in Biology

1669   **Marcello Malpighi** (Italian)—first to describe metamorphosis in the silkworm; discovered the excretory tubules of insects.

1737   **Jan Swammerdam** (Dutch)—prepared a detailed monograph on insect structure including many fine illustrations.

1824   **Straus-Durckheim** (German)—produced an illustrated work on insect anatomy: a detailed study of the European beetle, the cockchafer.

1834   **Leon Dufour** (French)—published several monographs describing the anatomy of several insect families.

1841   **George Newport** (English)—elucidated the embryology of insect phyla.

1864   **Franz Leydig** (German)—first to study tissues of insects under the microscope.

1959   **V. B. Wigglesworth** (English)—elucidated the physiological process that controls metamorphosis in insects.

1963   **G. G. Johnson** (English)—determined the means by which insect flight carries them over large distances.

1964   **Wolfgang Beerman** and **Ulrich Clever** (German)—discovered that chromosome puffs seen in insect species are active genes.

1965   **Miriam Rothschild** (English)—found the role of hormones in the life cycle of the rabbit flea.

1967   **Suzanne Batra** and **Lekh Batra** (American)—studied the mutualistic relationship between some insects and fungi.

1977   **G. Adrian Horridge** (English)—explained the workings of the ommatidia of the insect compound eye.

1978   **Lorus Milne** and **Margery Milne** (American)—presented a research study on the four types of insects that live on the surface of quiet waters.

1980   **William G. Eberhard** (American)—discovered the function of the horns in the horned beetle as organs of lifting.

1982   **Thomas D. Seeley** (American)—discovered how honeybees locate a site for a hive.

Words for Study

abdomen	green glands	oviduct
arthropod	hemocoel	ovipositor
book lungs	hemolymph	palp
cephalothorax	hormone	pericardial cavity
chelicerae	jointed appendages	polymorphism
compound eyes	juvenile hormone	prothoracic gland
corpus allatum	larva	pupa
crustacea	Malpighian tubules	spermatotheca
diapause	metamorphosis	spinneret
drones	nymphs	spiracles
ecdysone	ommatidia	swimmerets
gastric mill	open circulatory system	tracheal tubes
gill	ostia	tympanic
gizzard		workers

Questions for Review

PART A.   Completion. Write in the word that correctly completes each statement.

1. The animal phylum that is largest in number of species and number of individuals is the ..1..
2. The development of eggs without fertilization is known as ..2..
3. The body of a spider is divided into ..3.. and an abdomen.
4. Simple eyes are known as ..4..
5. Green glands have a (an) ..5.. function.

6. The most advanced class of the modern anthropods are the ..6..
7. The grasshopper has ..7.. mouthparts.
8. Wing stubs in mosquitoes are used for the purpose of ..8..
9. A hormone is secreted by ..9.. or ductless glands.
10. A membrane specialized for gathering sound vibrations is the ..10.. membrane.

PART B. Multiple Choice. Circle the letter of the item that correctly completes each statement.

1. The most outstanding characteristic of the anthropods is
   (a) membranous wings   (c) simple eye
   (b) bilobed antennae   (d) jointed appendages

2. Book lungs are the breathing mechanisms of
   (a) arachnids   (c) lobsters
   (b) termites   (d) grasshoppers

3. Examples of crustacea are
   (a) barnacles and beetles   (c) horseshoe crab and shrimp
   (b) lobsters and sow bugs   (d) water flea and mite

4. Ommatidia are structural and functional units of
   (a) thorax   (c) compound eye
   (b) flame cell   (d) brain

5. Lobster embryos
   (a) are free-swimming
   (b) fall to the bottom of the sea
   (c) are maintained in male the green gland
   (d) attach to the female swimmerets

6. The group that represents the dominant form of terrestrial life is the
   (a) insect   (c) scorpions
   (b) human   (d) spiders

7. The insect body is divided into the
   (a) cephalothorax and abdomen (c) head, tail, wing
   (b) wing, abdomen, head   (d) head, thorax, abdomen

8. The breathing holes on the abdomen of the grasshopper connect to
   (a) green glands   (c) Malpighian tubules
   (b) nephridia   (d) tracheal tubules

9. A nymph refers to a young
   (a) maggot   (c) honey bee
   (b) grasshopper   (d) blow fly

10. The growth and differentiation that takes place in insect larva is controlled by the hormone known as
    (a) juvenile   (c) ecdysone
    (b) brain   (d) prothoracic

PART C.   Modified True-False. If a statement is true, write "true" for your answer. If a statement is incorrect, change the underlined word to one that will make the statement true.

1. The body cavity of the arthropods is replaced by the <u>coelom</u>.

2. Arachnids are <u>ants</u>.

3. Malpighian tubules change nitrogenous wastes into dry crystals of <u>bile</u>.

4. <u>Spinnerets</u> circulate water over the gills of the lobster.

5. The lobster has a (an) <u>closed</u> circulatory system.

6. The only flying invertebrates are the <u>waterfleas</u>.

7. Butterfly mouth parts are structured for <u>biting</u>.

8. The feeding stage of the developing insect is the <u>pupa</u>.

9. Change in body form of the butterfly is known as <u>cocoon</u>.

10. Juvenile hormone is secreted by the <u>corpus allatum</u>.

Answers to Questions for Review

PART A

1. Arthropoda
2. parthenogenesis
3. cephalothorax
4. ocelli
5. excretory
6. insects
7. biting
8. balancing
9. endocrine
10. tympanic

PART B

1. d
2. a
3. b
4. c
5. d
6. a
7. d
8. d
9. b
10. c

PART C

1. hemocoel
2. spiders
3. uric acid
4. swimmerets
5. open
6. insects
7. sucking
8. larva
9. metamorphosis
10. true

CHAPTER 10

CHORDATES: HEMICHORDS TO MAMMALS

The chordates are animals that have a notochord at some time in their life cycle. The notochord is a living, internal skeletal axis in the form of a compact cellular rod-like structure that extends the length of the body. It lies dorsal to the digestive tract and is not to be confused with the backbone. In lower chordates (invertebrate chordates), the notochord prevents the body from shortening when muscles contract. During the embryonic stages of the vertebrate chordates, the notochord is replaced by a column of bones, the *vertebrae*, which form the backbone.

Chordates are set apart from lower animals by several distinguishing characteristics in addition to having a notochord. First, all chordate embryos have the three primary germ layers from which all specialized tissues and organs develop. Secondly, chordates are bilaterally symmetrical animals with anterior-posterior differentiation. Thirdly, the body has a true coelom and a digestive tract that begins with a mouth and ends with an anus. Other characteristics which differentiate the chordates from other animals are the presence of *pharyngeal gill slits* and the *dorsal hollow nerve cord*. The pharyngeal gill slits are paired vertical slits in the wall of the *pharynx* (throat) positioned directly behind the mouth and connecting to the outside of the body. The gill slits may be present only in the embryo stage, or they may persist and be used in breathing. The dorsal hollow nerve cord is a hollow cylindrical tube, usually expanded into a brain at the anterior end. It lies dorsal to the notochord. The hollow space is called the *neurocoel*.

The phylum Chordata is divided into four subphyla: Hemichordata, Urochordata, Chephalochordata, and the Vertebrata. The first three phyla represent the prevertebrates, also referred to as the invertebrate chordates.

The Prevertebrates

SUBPHYLUM HEMICHORDATA (ENTEROPNEUSTA)

Chordates in this group are worm-like animals of uncertain taxonomic position. They are marine, either living near the shore or in deep sea. Some forms are colonial. The representative organism is *Balanoglossus*, the acorn worm (Fig. 10.1).

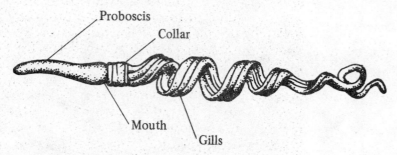

Fig. 10.1 *Balanoglossus*, the acorn worm

SUBPHYLUM UROCHORDATA—TUNICATA

The adult tunicates are sessile, sac-like animals, often referred to as *sea squirts*. They are filter-feeders, taking in water which passes through the gill slits into the atrial chamber and out by way of the *atriopore*. The tunicates reproduce asexually by budding and also sexually by eggs and sperm. The larval forms have a notochord, a nerve cord, a pharynx with gill slits, and an *endostyle*. An endostyle is aciliated or glandular out-pocketing from the wall of the throat of the urochords. Cilia and mucus from the gland sweep food backwards into the gullet. The adult forms lose the chordate characteristics. Examples of the tunicates are *Cynthia*, *Salpa* and Sea Pork.

SUBPHYLUM CEPHALOCHORDA—LANCELETS

The representative organism for the group is *Amphioxus*, the lancelet. It burrows in the sand at the shoreline of tropical or temperate waters. The body is small, elongated, and shaped like a fish without paired fins. It has gill slits, a well-developed notochord and a dorsal hollow nerve cord (Fig. 10.2).

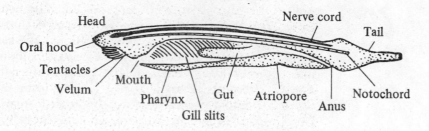

Fig. 10.2 *Amphioxus*

The Vertebrates

Animals that have a true backbone composed of segmented parts called *vertebrae* belong to the chordate subphylum Vertebrata. The vertebrae may be made of cartilage or bone: if made of the latter, cartilage cushions prevent the bones from rubbing together. The backbone is built around the notochord and usually obliterates it. Vertebrates vary in size from large to small, but all have a living endoskeleton usually made of bone. A limited number of water-dwelling species exhibit an endoskeleton made of cartilage. All vertebrate species have marked development of the head where a brain is enclosed in a *cranium*.

Blood is pumped through a *closed circulatory system* by means of a ventral heart, having at least two chambers: an *atrium* and a *ventricle*. The *hepatic portal system* carries blood laden with food from the intestines to the liver before it reaches the body cells. Vertebrate red blood cells contain the iron-bearing pigment hemoglobin which is specialized to carry oxygen. Such a system of closed blood vessels prevents blood from entering the body cavity.

Most vertebrates (except human) have a post-anal tail which is a continuation of the vertebral column. Although there are never more than two sets of paired appendages, some adult vertebrates show only one such set or none at all, the appendages having been lost over evolutionary time. Evidence of lost appendages may be seen in embryonic forms or may be demonstrated by *vestigial* structures. The coccyx bone in humans is a remnant (vestigial structure) of a post-anal tail. Other characteristics of vertebrates include a mouth that is closed by a movable lower jaw and a thyroid gland derived from the ventral wall of the pharynx. In the invertebrate chordates the endostyle is an evolutionary signpost pointing to the development of the thyroid gland.

CLASS AGNATHA

The Agnatha are the most primitive of the vertebrates. These are the jawless fishes—lampreys and hagfish—characterized by a round mouth that has earned them the general name *cyclostome*. The agnathans have only a single nasal opening at the tip of the snout and no paired fins or paired limbs of any kind. The notochord is present throughout life and the skeletal structures are made of cartilage. The skin is smooth and slimy, lacking scales. The eyes are rudimentary.

The lampreys (Fig. 10.3) live in freshwater where they are parasite-predators on bony fish. The mouth of the lamprey is a sucker disk used by the fish to attach itself to rocks or other organisms. After attaching itself to a fish, the lamprey uses a rasping tongue-like structure to break the skin and suck the blood of the host. The lampreys produce an *amnocoete larva*, a filter-feeder living buried in river mud, which lasts for about seven years before changing to adult form. The adult life span is short, the adults dying after going upstream to spawn.

Hagfish live in temperate and tropical marine waters where they are scavengers, feeding on dead, disabled, and diseased fish.

TABLE 10.1.   Differences in Structure Between the Invertebrates and the Vertebrates

	Invertebrate	Vertebrate
skeleton	nonliving exoskeleton	living endoskeleton
nerve cord	ventral, double and solid; formed by delamination* from ectoderm.	dorsal, single and hollow; formed by invagination of** ectoderm.
heart	dorsal	ventral
hemoglobin	in plasma, when present	in blood cells
circulatory system	open with hemocoel and sinuses	closed, contained in vessels

* Delamination means splitting off from ectoderm
** Invagination refers to an inpocketing

Fig. 10.3   Lamprey

CLASS CHONDRICHTHYES — CARTILAGINOUS FISH

The sharks (including the dogfish), rays and skates are examples of the Chondrichthyes. Most species in this class live in the ocean. The Chondrichthyes have paired fins, a lower jaw, and gill arches. The body of the shark is covered with *placoid scales* which arise from the ectoderm, also forming the teeth in the jaws and on the roof of the mouth. A distinctive feature of this group of fish is that the skeleton is made of cartilage, not bone as in the Osteichthyes, or bony fish. The sharks and their relatives differ from the bony fish in other ways as well: they have no swim bladder, no true scales and the gill slits are uncovered.

Sharks are predators and feed upon bony fish. Most of the digestion takes place in the stomach while absorption of digested food occurs in the intestine. A *spiral valve* that extends the length of the short, fat intestine increases surface area for absorption. The mouth is located ventrally (Fig. 10.4).

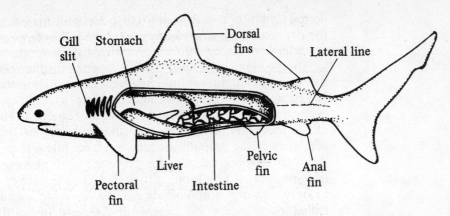

Fig. 10.4 Structure of the shark

Of particular interest is the sensory system of the shark. The eye of the shark, except for shape, size and absence of eyelids, is very much like the human eye. The paired nostrils are pits on the ventral surface of the head that open externally only and do not empty into the throat. The nostrils are lined with an olfactory membrane which is connected by nerve fibers to the olfactory lobes in the brain. Along the sides of the body are pressure receptors known as *lateral line systems*. In the head pits of the lateral line organs have become modified into long canals filled with mucous. These canals, known as *ampullae of Lorenzini*, are sensitive electrochemical receptors.

In all cartilaginous fish, fertilization is internal with the male using *claspers*, modified pelvic fins, to place the sperm in the female's body. In some species the embryos are nourished *ovoviviparously*, obtaining food from the egg. In other species, the embryo is maintained inside the mother's body and nourishment is passed from the blood vessels of the mother into the blood system of the developing embryo; this method of embryo nutrition is known as *viviparous*.

SUPERCLASS PISCES

Superclass Pisces includes all of the bony fish. Like the cartilaginous fish, the Pisces are water-dwelling vertebrates that breathe by means of gills, have paired eyes, and a two-chambered heart with blood flowing from the heart through the gills and then to the other parts of the body. The Pisces differ from cartilaginous fish in having *dermal* scales and bone, not cartilage, in the skeleton.

Major Groups

The superclass Pisces is divided into four subgroups: classes Dipnoi, Crossopterygii, Ganoidei, and Osteichthyes.

The Dipnoi are lungfish that live in seasonally dry estuaries in Africa, Australia and South America. These fish breathe by means of lungs and by gills. The lung allows these animals to obtain oxygen from the air and permits them to live in water that is too foul for gill breathing. During the dry season, they aestivate in dry mud; they become activated when the river bed fills up with water during the rainy season.

Some biologists consider the lungfish to be intermediate between the fish and the amphibia. Lungfish have several body structures that look like those found in primitive land dwellers, and the early part of the life cycle of the lungfish and the frog are almost identical. However, the shape of the teeth and jaws are not froglike and indicate that direct line descendency of the amphibians may not have taken place.

The crossopterygians, for the most part, represent fish that have become extinct. This group is thought to be the ancestral line from which modern fish and amphibia descended. A living member of this group called a *coelocanth* was caught off the coast of South Africa. It was a large, lobe-finned fish with bluish scales.

Members of the Ganoidei are the most primitive of the bony fish. Most species in this class live in fresh water and their distinctive feature is the heavy armor of *ganoid* scales which are flat and form heavy plates. Once the armored fish were the dominant form of fish, but now only a few scattered species remain. Living examples of the Ganoidei are sturgeon, gar, pike and freshwater dogfish. Some species have partly cartilaginous skeletons and a degenerating spiral valve indicating possible relationship to the sharks.

The Osteichthyes, or true Bony Fish, are the most highly organized of all the fish. They belong to the order Teleostei and are referred to as *teleosts*. They are a very successful and widely distributed group, differing widely in aquatic habitat, size, feeding patterns, and shape. Fig. 10.5 illustrates the physical diversity among the teleosts. However, there are some characteristics that are typical of all species of bony fish.

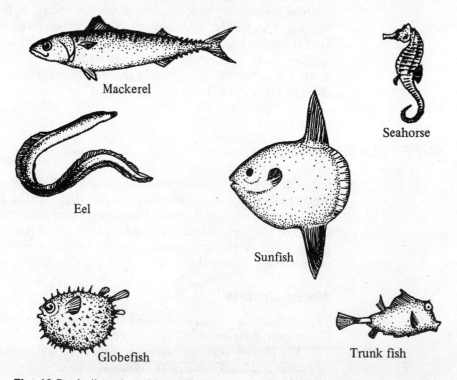

Fig. 10.5 A diversity of bony fishes

Distinctive Features

The distinctive features of the teleosts that make them different from the cartilaginous fish are primarily the skeleton, the scales and the gills. The internal skeleton is made of bone. The scales are either *cycloid* (smooth) or *ctenoid* (rough) and are constructed as thin bony plates rather than like the thick plates of the ganoid type. The gills are reduced in number to four pairs, instead of the seven pairs in sharks, and are covered with a flap of bone called the *operculum*. The fins are paired. Fig. 10.6 shows the location of the dorsal, anal, pelvic and pectoral fins. The *swim bladder* is an outgrowth of the pharynx, an oxygen-filled structure that enables the fish to float.

Some sense organs in bony fish are well developed; others are not. The lateral line receptors are sensitive to movements of current. The olfactory pits are not connected to the mouth and therefore have no respiratory function, but they do enable fish to respond to chemical stimuli. Eyes of fish vary in size according to species. The taste buds located in the mucosa of the mouth are poorly developed. Fish probably cannot hear in air. There are three semicircular canals that control equilibrium, but auditory vibrations reach the ear through the skull bones.

Reproductive Behavior

Patterns of reproduction vary among the bony fish, but in most species fertilization is external. During certain breeding seasons the female lays thousands of eggs known as *roe*. The male deposits sperm cells known as *milt* close to the eggs. The sperm, attracted by some chemical substance given off by the eggs, swim to them. One sperm enters an egg and fertilizes it. The fertilized egg (zygote) then goes through a series of mitotic divisions (cleavage), resulting in a new immature fish called a *fry*. The fry has attached to it a sac appropriately called the *yolk sac* because it contains yolk which nourishes the young fish until it can feed independently (Fig. 10.7).

Fertilization in fish is chancy: many sperm never reach the eggs and many fertilized eggs die before development. Hence there is an overproduction of gametes to ensure the survival of the species.

In a few fish species fertilization is internal and in a few parental care is given to the fertilized eggs. The stickleback male, for example, takes care of the fertilized eggs in nests and the male seahorse carries them around in a brood pouch.

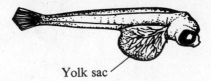

Yolk sac

Fig. 10.7 An immature fish

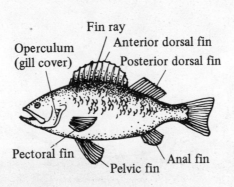

Fin ray
Operculum (gill cover)
Anterior dorsal fin
Posterior dorsal fin
Pectoral fin
Pelvic fin
Anal fin

Fig. 10.6 The bony fish

CLASS AMPHIBIA—FROGS AND THEIR RELATIVES

From an evolutionary perspective, amphibia are transitional animals, living first in water and then on land during specific times in the life cycle. They are equipped for this double life, so to speak, by body structures and organs that change to meet their needs.

An amphibian must spend part of its life cycle in the water where its eggs are laid and fertilized. The eggs develop into a larval stage, or tadpole, that has fish-like characteristics. In tadpoles breathing is by means of gills, blood is pumped by a two-chambered heart, and swimming is by means of a tail and body movements made possible by muscles in the body wall. The change to adult form is known as *metamorphosis*, a process controlled by the thyroid gland. The adult amphibian loses the gills, lateral line senses, tail, unpaired fins and muscles controlling them—the fish characteristics—and develops structures adapted for life on land. An adult amphibian breathes by means of lungs and has a three-chambered heart which is more efficient at pumping blood between the lungs, the heart, and the rest of the body. In most species the adult also has limbs for movement, but no tail (Fig. 10.8).

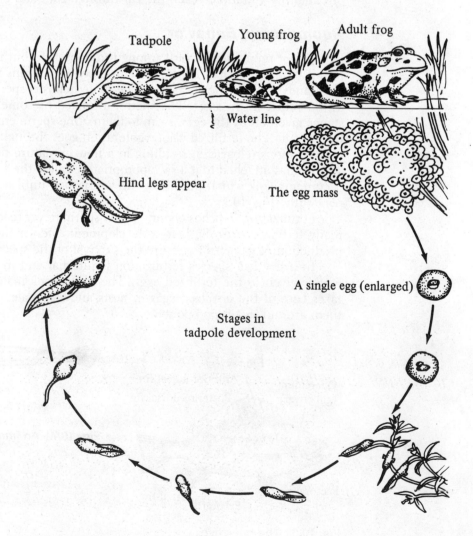

Tadpole   Young frog   Adult frog

Water line

Hind legs appear

The egg mass

A single egg (enlarged)

Stages in
tadpole development

Fig. 10.8   Metamorphosis in the frog

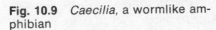

Fig. 10.9   *Caecilia*, a wormlike amphibian

Fig. 10.10   *Necturus*, the mud puppy

There are three general types of amphibia. The Apoda are wormlike, legless, ground-burrowing forms found in tropical and semitropical regions; *Caecilia* (Fig. 10.9) is a representative genus. The Urodela are amphibians that do not lose their tadpole-like tail in metamorphosis; salamanders, newts and mud puppies (Fig. 10.10) are examples. The Anura, represented by frogs and toads, lose their tails on becoming adults. Frogs and toads are the first vertebrates to become vocal.

In frogs fertilization is external. Sperm leave the testes through tubules called *vasa efferentia* which communicate with the kidney. The sperm cells then pass into the *Wolffian duct* which leads to the *cloaca*, a passageway that opens to the outside of the body. In the female large egg masses are released into the body cavity from two ovaries, located at the anterior end of each kidney. Beating cilia sweep the eggs into coiled tubules known as *oviducts* where they are propelled to the cloaca and then out of the body. As the eggs pass through the oviducts they are coated with a thin layer of jelly-like material. At the time when the female is depositing eggs in the shallow waters of a pond or brook, the male deposits sperm over them. The sperm swim to the eggs and as each sperm reaches an egg, it digests its way through the jelly and into the egg, effecting fertilization. After fertilization the jelly coating on the eggs swells due to the absorption of large amounts of water. The swelling of the black jelly causes the eggs to adhere together and protects them from predation by fish and other animals. The fertilized egg, or zygote, undergoes cleavage, forming a tadpole.

Amphibians demonstrate some interesting mechanisms for prolonging life. Many have powers of regeneration to the extent that entire organs can grow back after having been lost. The axolotyl and the urodele (*Triton cristatus*) are examples of animals in which regeneration occurs. Many amphibians are able to hibernate or aestivate and survive unfavorable conditions.

CLASS REPTILIA

Reptiles are the first true land vertebrates relieved of the necessity of returning to the water to reproduce. Reproducing on land was made possible by the development of special embryonic membranes that preserve on land the protection offered by the aquatic environment to the embryos of lower vertebrates. One such membrane is the *amnion*, a membranous sac surrounding the embryo and filled with water. This sac of water prevents the delicate, rapidly dividing embryonic cells from drying out and protects them from shock and mechanical injury. Animals having the amnion are called *amniotes*; those animals without an amnion are known as *anamniotes*. Reptiles, birds and mammals are amniotes and therefore adapted to life on land.

The eggs of reptiles are fertilized internally, and the female lays fertilized eggs. Reptile embryos develop encased in an egg surrounded by a leathery shell. Lining the shell is the *chorion*, an embryonic membrane that mediates the two-way exchange of gases between the outside air and the embryo. Serving as a temporary organ of respiration and excretion is a third embryonic membrane called the *allantois*. The allantois, bearing a capillary network, grows into the space between the chorion and the amnion.

General Characteristics

Reptiles have a dry leathery skin covered with epidermal scales. A somewhat flattened skull contains a brain having a cerebrum much larger than that of the fish or amphibians. The eyes have secreting glands which keep the surface moist. Some species of reptiles have an external pore (*meatus*) positioned on the side of the head connected to the auditory ossicle or bone in the middle ear. Reptiles are air-breathers and have rather well-developed lungs. The heart is composed of two atria and a ventricle; in some species the ventricle is almost divided into two compartments, an evolutionary signpost pointing to the four-chambered heart. The body temperature of reptiles is not constant, changing with the external environment. In popular speech, such animals are called cold-blooded; in technical language, *poikilotherms*.

Major Groups

Extinct Forms

Reptiles flourished during the Mesozoic Era which lasted for about 130 million years and came to an end about 65 million years ago. At that time there were more than twelve orders of reptiles. Today only four orders of reptiles remain. Among the ancient reptiles there was a great diversity of form and adaptations that permitted life in a variety of habitats: dry land, water, swamps, and air. The *stem reptiles* had many features in common with primitive amphibians and resembled modern-day lizards externally. The *therapsids*, however, were more advanced, having teeth which looked much like those of the mammals. The dinosaurs are remembered for being very large and yet some were quite small. It is believed that many of the dinosaur species were able to walk on two legs, using the enormous tail for balance. The shorter front legs were probably used when walking slowly or resting. Some of the larger dinosaurs returned to the sea. The *icthyosaurs* were best adapted for aquatic life. A few of the ancient reptiles were able to fly; the *pterosaurs* had wings but they probably did more gliding than true flying.

Living Forns

The living reptiles are classified in four orders: Rhynchocephalia, Chelonia, Squamata and Crocodilia.

Order Rhynchocephalia—Tuatara   The only living representative of this order is the New Zealand "tuatara" of the genus *Sphenodon*. This lizard is about one and a half meters long having a median eyestalk at the top of the head. *Sphenodon* has a number of primitive reptilian characteristics, including no tear glands, teeth in the roof of the mouth, a lung that resembles a cluster of toad's lungs, abdominal ribs, unfused frontal skull bones, and a vertebral column with a fishlike structure.

Fig. 10.11   Turtle

Fig. 10.12   Crocodile

Order Chelonia—Turtles and Tortoises   The chelonians are the turtles and tortoises. The skeleton is modified to form a box-like covering, the upper curved portion of which is called the *carapace*, the lower part, the *plastron*. The head and the tail are the only movable parts of the animal. The jaws are horny and toothless. Chelonians live on land, in freshwater, and in the sea. More turtles live on the American continent than anywhere else (Fig. 10.11).

Order Squamata—Lizards and Snakes   Lizards and snakes are squamates. Their bodies are covered with a great number of small flexible scales that cannot be removed easily like the scales of bony fish. Lizards have movable eyelids, visible earpits and usually have legs. Snakes, on the other hand, are legless, and do not have eyelids or earpits.

Lizards are typically land dwellers requiring the warmth of the sunshine. The iguana is a large tree-living lizard native to Mexico. The chameleon, native to Africa, has a flexible grasping tail and grasping feet; its coat is capable of changing colors to match the background. The gila monster *Heloderna* is the only lizard whose bite is poisonous.

Although snakes are cold to the touch, they are not slimy. The body is heavily muscled and strongly ribbed. Snakes use the ribs in walking. Of the 110 species of snakes in the United States only 20 species are poisonous. Among the poisonous snakes are the copperhead, rattlesnake, water moccasin, and the coral snake. The harmless varieties include the puff adder, the garter snake, the black snake and the milk snake.

Order Crocodilia—Crocodiles and Alligators   This order includes the crocodiles and the alligators, large thick-bodied reptiles. These are considered to be the most advanced of the reptiles. The heart has a ventricle that is almost completely divided into two compartments. The lungs are very well developed and the brain has a rather large cerebrum. They live in shallow salt or fresh water where they are better able to locomote than on land (Fig. 10.12).

CLASS AVES—BIRDS

Birds are terrestrial vertebrates with feathers. Feathers are the distinctive feature of birds: all birds have them and no other animals are so covered. The forelimb is modified into wings for flight, leaving the hindlimbs for walking (bipedal locomotion). Birds are built for flight; special adaptations in body structure effect lightness in weight, efficiency and strength. Not only are the feathers light in weight and easily moved and lifted by wind, but they also create warmth next to the body. Body heat warms the air that is in contact with the bird's body. Warm air becomes lighter and rises. Other adaptations for flight are the compact, but hollow, bones, numerous air sacs occupying all available body spaces, reduced rectum, loss of teeth, and tail feathers replacing a bony tail (Fig. 10.13).

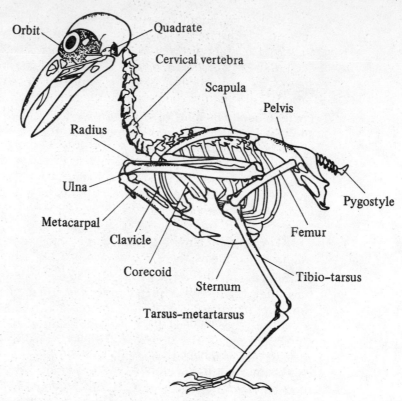

Fig. 10.13 Bone structure of the bird

In most bird species, the wings are organs of flight. However, in some species, the wings are modified for other purposes. For example: the wings of penguins serve as flippers for swimming.

According to fossil evidence, the evolutionary link between the reptiles and the birds was *Archaeopteryx*. Biologists consider *Archaeopteryx* to have been a bird because it had feathers, but it also had some very distinctive reptilian features, such as teeth and a long, bony tail.

Modern birds have lost reptilian characteristics. The feet of birds of various species are modified for all kinds of uses: running, scratching, grasping, swimming and wading. The horny bill is a characteristic of all birds. It is toothless and modified in various shapes to carry out special functions. The colors of bird feathers are due to pigments or irridescence and indicate *sexual dimorphism*, a state in which the male is often brightly and elaborately colored while the female is drab.

The brain of the bird shows greater development than the brain of the reptile, and the heart has four chambers—two atria and two ventricles, permitting complete separation of oxygenated and deoxygenated blood. Unlike the reptiles, birds maintain constant body temperature: they are *homiotherms*, or warm-blooded. The digestive system consists of a crop, a stomach, a gizzard, an intestine and a much reduced rectum. Birds do not retain their feces, a condition which is an adaptation for flight. There is a well developed kidney that empties by way of the ureter. Birds have remarkable eyesight, a feature that makes hawks and eagles such successful predators. On the other hand, the sense of smell is very poorly developed in birds. The *nictitating membrane* (third eyelid) is a translucent membrane that covers the eye crosswise serving as protection during flight.

Birds show another advancement over reptiles in behavior. Song is used to establish territorial rights, an important aspect of bird social behavior. The song of birds is produced by air passing over the *syrinx*, a secondary larynx located at the lower end of the windpipe (trachea) at its junction with the bronchi. The pattern of reproduction in birds also differs from that of reptiles and amphibians. Birds practice monogamy, the mating of one male with one female either for life or for the duration of the breeding season. Before mating birds go through courtship, a special type of behavior engaged in by males trying to gain the attention of reproductive females. In birds, fertilization is *internal*; the development of the young, external. In most species the female lays the eggs in a characteristic-type nest and both the male and the female take turns *setting*, a process of incubation by which the eggs are kept warm, and later caring for the young.

The testes of birds are positioned in the back just above the kidneys. Mature sperm leave the testes through tubules called the *ductus deferens* which lead into the cloaca. The gonads of females are the ovaries, glands where eggs are produced. In female birds, there is one functioning ovary; the other having degenerated early in the bird's life. A bird's egg cell consists of a nucleus and a little protoplasm. The cell is surrounded by yolk. A mature cell and its yolk is drawn into the upper end of the oviduct which is funnel-shaped and contains waving cilia. As the cell with its yolk travels through the oviduct, it is surrounded with layers of the protein *albumen*. The albumen is the white of the egg. The albumen is surrounded by a thin membrane which then is covered by a calcareous shell secreted by lime-producing glands that line the lower end of the oviduct (Fig. 10.14).

Fertilization is accomplished during mating at which time the male and female place their cloacas close together. Sperm swim from the cloaca of the male into the female cloaca and up into the oviduct. Fertilization takes place high up in the oviduct before the albumen and the other surrounding membranes are secreted by the oviduct cells. Most birds lay of *clutch* of less than six eggs. However, ducks may lay as many as 15 eggs at one time (Fig. 10.15).

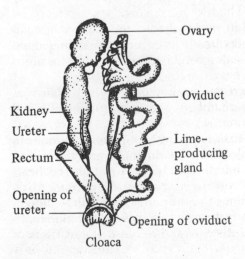

Fig. 10.14 Reproductive system of the female bird

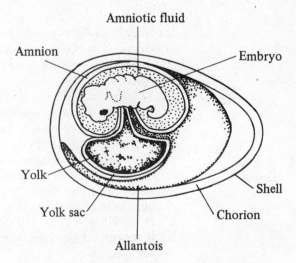

Fig. 10.15 Development of the bird embryo

CLASS MAMMALIA

The first mammals appeared on earth about 65 million years ago. The early mammals were small and insignificant, but had developed adaptations enabling them to survive at the same time that the giant reptiles roamed the earth. It is believed that the early mammals were no larger than field mice, feeding on insects and scraps of vegetation. The large eye sockets of the fossil forms indicate that these primitive mammals were nocturnal and probably lived in trees.

Scientists present various reasons for the success of the mammals. First among these is the fact that they were fast movers and their food requirements and behavioral patterns kept them out of the way of carnivorous (flesh-eating) reptiles. Second, the early mammals have been accused of contributing to the demise of the dinosaurs by eating the unprotected eggs of these reptiles. Third, the change in form of the skeleton which allowed the positioning of relatively thin legs underneath the animal led to a narrow walking track and contributed to faster movement. (Remember how the thick legs of a crocodile are spread out from the body on a wide track.) Fourth, the refinement of the lower jaw—movable and composed of a single bone—coupled with the development of different kinds of teeth for biting, tearing and chewing made the mammals successful carnivores.

Modern mammals show a great deal of diversity. They vary in size from a shrew hardly more than 2.5 centimeters in length to the enormous blue whale which may measure more than 30 meters in length. Mammals are widely distributed over the earth and are adapted to live in diverse habitats on land (wolves), in water (sea lions), in underground burrows (gophers), in desert burrows (kangaroo rat), in open seas (whales), in the air (bats) and in forests (baboons).

General Characteristics of Mammals

The characteristics that set mammals apart from other animals and made them adaptable to a wide range of habitats are as follows:

1. Mammals have *mammary glands* (from whence the name mammal) that supply the young with milk directly after birth. Newborn mammals are usually quite helpless and depend upon the mother for nourishment.
2. At some time during the life cycle, all mammals have *hair*. Hair is as typical to mammals as feathers are to birds and scales to bony fish and reptiles.
3. Mammals are warm-blooded. Constant body temperature is due, in part, to the four-chambered heart, a device which prevents the mixing of oxygenated and deoxygenated blood. The four-chambered heart first appears in birds. However, in mammals there is another characteristic of blood that contributes mightily to stable body temperature. The red blood cells on maturity lose their nuclei and mitochondria, restricting them to function as oxygen carriers, not as users of oxygen.
4. Most species of mammals have *sweat glands* which provide a secondary means of excreting water and salts.

5. Mammalian teeth have evolved into three different types: incisors for tearing; canines for biting; molars and premolars for grinding.

6. All but a few species have seven vertebrae in the neck. These neck bones are known as *cervical vertebrae*.

7. A muscular diaphragm separates the thoracic cavity (containing the lungs and the heart) from the abdominal cavity (housing part of the digestive system, the reproductive organs and the excretory system).

Major Groups

The class Mammalia is divided into three major groups: Protheria, Metatheria, and Eutheria.

Subclass Prototheria—Monotremes

The subclass Prototheria consists of a single order, Monotremata. The monotremes are primitive egg-laying mammals. The eggs, large and full of yolk, house the developing monotreme embryos. Examples of the monotremes are the "duckbill" or platypus (*Ornithorhynchus*) indigenous to Australia and Tasmania (Fig. 10.16); the spiny anteater (*Echidna*), also an inhabitant of Australia; and a long snouted anteater (*Proechidna*) indigenous to New Guinea. Modified sweat glands of the anteater secrete a milk substitute which the young lick up from tufts of hair on the mother's belly.

Subclass Metatheria—Marsupials

The subclass Metatheria consists of a single order, *Marsupialia*. The marsupials are primitive mammals that do not have a placenta. The young are about 5 centimeters long at birth and are in an extremely immature condition. At birth they crawl into the mother's pouch or *marsupium*. The rounded mouth is attached to a nipple and the mother expresses milk down the throat of the helpless fetus. As development occurs, the young marsupial is then able to obtain milk by sucking. There are 29 living genera of marsupials, 28 of which live in Australia. The oppossum *Didelphys* is indigenous to North, South and Central Americas and *Caenolestes* inhabits regions of Central America only. Besides the oppossum, other marsupials are the kangaroo, koala bear, Tasmanian wolf, wombat, wallaby and native cat. It is believed that at one time a land bridge connected South America to Australia. With the disappearance of this land link, Australia became geographically isolated, permitting the existence of marsupials that do not have to compete with more advanced mammals.

Subclass Eutheria—Placental Mammals

The subclass Eutheria includes all of the modern mammals that have a placenta. You will recall that in the reptile and the bird, a membranous sac called the allantois serves as an organ of respiration and excretion for the developing embryo. The reptile and bird embryos develop outside the mother's body and are nourished on stored yolk in the egg. The eutherians develop inside of the mother's body in a muscular sac, the *uterus*. The region where the allantois comes in contact with the uterine wall is heavily supplied with capillaries, the smallest blood vessels in the body. At this particular site the *placenta* is formed. The embryo is attached to the placenta by an *umbilical cord* which contains a fetal artery

Fig. 10.16 Duckbill platypus

and a fetal vein. By diffusion from the capillaries of the mother, food nutrients and oxygen travel into the capillaries in the embryonic placenta. The fetal artery collects blood from the capillaries of the placenta and conducts it to the capillaries of the embryo. The fetal vein collects the blood of the embryo laden with the wastes of fetal respiration and conducts it away from the embryo. It is important to emphasize that the blood of the mother and the blood of the embryo do not mix. Gasses and nutrients from the maternal circulation pass to the fetal circulation by diffusion across capillary walls.

Table 10.2 provides a quick summary of some of the placental mammals.

TABLE 10.2.   Some Placental Mammals

Orders	Characteristics	Examples
Insectivora	Small; nocturnal; burrowing or tree-living; feed on insects; sharp-snouted.	Hedgehog, moles, shrews
Dermoptera	Link between the insectivores and the bats.	Flying lemur
Chiroptera	Winged, only mammals able to fly; nocturnal; identify objects by echolocation; according to genus feed on insects, blood, fruit.	Bats
Carnivora	Predators; swift of foot, collar bones reduced, teeth specialized for meat-eating; cerebral development.	Dogs, wolves, coyotes, bears, cats, lions, cheetah, foxes, raccoons, weasels, skunks, etc.
Rodentia	Gnawing mammals; sharp incisor teeth; plant eaters; most numerous of all living mammals.	Hares, squirrels, guinea pigs, rats and mice, beavers, muskrats, porcupines, prairie dogs, woodchucks
Primates	Evolved from tree-dwelling Insectivora ancestors; most species arboreal; teeth unspecialized; marked development of eyes and brain, especially the cerebrum; quadrupeds, but upright sitting posture.	
Lemuroidea	Tree-dwelling; primitive; small, pointed ears; long snout; big toes and thumbs set apart from other digits.	Lemurs
Tarsioidea	Binocular vision; reduced sense of smell.	Tarsiers
Anthropoidea	Enlarged, convoluted cerebrum; well-developed eyes; prehensile tails; broad, flat noses; thumbs reduced; external nostrils close together.	Monkeys, baboons, macaques, gibbons, orangutans, gorillas, chimpanzees, humans

Chronology of Famous Names in Biology

1750   **Rene Antoino deReaumur** (French)—wrote a monumental work in six volumes on the anatomical structure of insects.

1757   **Pierre Lyonet** (Dutch)—presented a brilliant piece of research on the life cycle of the goat-moth caterpillar.

1760   **Petrus Camper** (Dutch)—published anatomical studies on elephant, rhinoceros and the reindeer. Published an excellent paper on the anatomy of the orangutan.

1763   **John Hunter** (English)—introduced "modern" methods in the study of comparative anatomy of vertebrates.

1804   **Alexander Brongniart** (French)—first to classify the amphibia separate from the reptiles.

1830   **Johannes Peter Muller** (Germany)—devoted himself to marine research and made valuable contributions to the study of the evolution of marine forms.

1837   **Karl Ernst von Baer** (Germany)—discovered that animal tissues arise from three primary germ layers.

1876   **Max Furbringer** (Germany)—produced valuable information on the comparative anatomy of the breast, wing, and shoulder of birds.

1957   **James Gray** (England)—carried out research which elucidated the mechanisms by which small fish gain swimming speed.

1965   **Archie Carr** (English)—studied the navigational habits of the green turtle.

1967   **Neal Griffith Smith** (England)—discovered that gulls recognize species mates by visual signals.

1968   **Kjell Johansen** (Swedish)—presented a remarkable study on the evolution of air-breathing fishes.

1969   **Crawford H. Greenewalt** (American)—elucidated the mechanical means by which birds produce song.

1971   **Knut Schmidt-Nielsen** (American)—discovered the pathway that air takes through the lungs of birds.

1977   **T. J. Dauson** (English)—studied the adaptive strategies of kangaroos that enable them to maintain their reproductive potential.

Words for Study

albumen	ganoid	placoid scales
allantois	hepatic portal system	plastron
amnion	lateral line	poikilotherm
ampullae of Lorenzini	mammary gland	roe
atrium	marsupial	spiral valve
auricle	metamorphosis	swim bladder
bipedal locomotion	milt	sexual dimorphism
carapace	monotreme	syrinx
chordate	neurocoel	teleost
chorion	nictitating membrane	umbilical cord
cloaca	notochord	uterus
clutch	operculum	vasa efferentia
cranium	oviduct	ventricle
ctenoid	oviparous	vertebrae
cyclostome	ovoviviparous	vestigial
dorsal nerve cord	pharyngeal gill slits	viviparous
ductus deferens	pharynx	Wolffian duct
endostyle	placenta	yolk sac
fry		

Questions for Review

PART A.   **Completion.** Write in the word that correctly completes each
statement.

1. The compact cellular rod that extends for the length of the chordate
   body is the ..1..
2. The type of symmetry exhibited by chordates is ..2..
3. The bones that compose the backbone are the ..3..
4. A structure that has lost its use over evolutionary time is called ..4..
5. The hepatic portal system carries blood from the intestine to the ..5..
6. The most primitive of the vertebrates are the ..6.. fish because of
   their feeding patterns.
7. Hagfish are best described as ..7..
8. The lamprey has ..8.. paired fins.
9. The amnocoete lives buried in ..9..
10. The type of scales that cover shark's body are ..10..
11. The pressure receptors along the sides of fish are known as ..11..
    systems.

12. A living "fossil" fish is the ..12..
13. The teleosts are the ..13.. fish.
14. The larval stage of the frog is the ..14..
15. The first vertebrates to become vocal are the ..15..
16. The adaptation that protects the embryos of terrestrial animals against drying and dessication is the ..16..
17. Dinosaurs are best classified as ..17..
18. The carapace and the plastron are best associated with the ..18..
19. All birds have ..19..
20. An example of a monotreme is the ..20..
21. An immature fish is called a ..21..
22. Egg masses of frogs leave the body cavity through tubes known as ..22..
23. Before mating birds go through ..23.. behavior.
24. The sperm of fish are known as ..24..
25. The marsupial young are born immature because there has been no development of the ..25..

PART B. **Multiple Choice.** Circle the letter of the item that correctly completes each statement.

1. The hollow space in the notochord is the
   (a) coelom   (c) neurocoel
   (b) hemicol   (d) coelenteron

2. The representative organism for the lancelets is
   (a) amphioxus   (c) acorn worm
   (b) cynthia   (d) sea pork

3. The position of the vertebrate heart is
   (a) dorsal   (c) ventral
   (b) lateral   (d) distal

4. The endostyle of the prevertebrates is an evolutionary signpost of the
   (a) brain   (c) gullet
   (b) pharynx   (d) thyroid gland

5. The hemoglobin in the vertebrate's blood system is contained in
   (a) vessels   (c) plasma
   (b) cells   (d) sinuses

6. The function of the spiral valve is to
   (a) increase surface area   (c) circulate blood
   (b) digest food   (d) store wastes

7. The process in which embryos obtain nourishment from the yolk in the egg is described as being
   (a) viviparous   (c) ammocoete
   (b) ovoviviparous   (d) osmosis

8. The dogfish is a (an)
   (a) amphibian   (c) bony fish
   (b) lamprey   (d) shark

9. Dipnoi is the class of the
   (a) sharks
   (b) skates
   (c) lungfish
   (d) cyclostomes

10. Teleosts are the
    (a) sharks
    (b) lung fish
    (c) lamprey
    (d) bony fish

11. The gill covering of the bony fish is the
    (a) spiral valve
    (b) operculum
    (c) cranium
    (d) chorion

12. A true statement about fish is that they
    (a) have well-developed sense organs
    (b) usually are farsighted
    (c) have three semicircular canals
    (d) have mouth-connected olfactory pits

13. The urodeles
    (a) give birth to live young
    (b) have tails
    (c) are usually legless
    (d) represent the reptiles

14. The fetal membrane that functions as an organ of respiration and excretion is the
    (a) amnion
    (b) allantois
    (c) chorion
    (d) shell

15. When an organ or structure "bears a capillary network", it
    (a) has hair
    (b) is divided
    (c) becomes septate
    (d) has blood vessels

16. The reptiles are true land animals because they
    (a) breathe free air
    (b) have strong walking legs
    (c) reproduce on land
    (d) feed in vegetation

17. Bipedal locomotion is a characteristic of
    (a) toads
    (b) turtles
    (c) crocodiles
    (d) birds

18. Hollow bones, numerous air sacs and loss of teeth are adaptations for
    (a) floating
    (b) swimming
    (c) diving
    (d) flying

19. The four chambered heart first appears in
    (a) mammals
    (b) birds
    (c) reptiles
    (d) amphibians

20. Metamorphosis is best associated with
    (a) fish
    (b) rabbits
    (c) frogs
    (d) strawberries

21. Roe and milt are best associated with
    (a) fish
    (b) frogs
    (c) starfish
    (d) jellyfish

22. The number of sperm cells necessary to fertilize an egg is (are)
    (a) one
    (b) two
    (c) three
    (d) four

23. The correct number of ovaries in the female bird is shown at the letter
    (a) one   (c) three
    (b) two   (d) four

24. Egg-laying mammals are known as
    (a) marsupials   (c) monotremes
    (b) tadpoles   (d) placentals

25. The production of song in birds is associated with the
    (a) nictitating membrane   (c) gizzard
    (b) syrinx   (d) bill

PART C Modified True-False. If a statement is true, write "true" for your answer. If a statement is incorrect, change the underlined word to one that will make the statement true.

1. The anterior end of the dorsal hollow nerve cord is usually expanded into a <u>throat</u>.

2. A ventral solid nerve cord is characteristic of <u>vertebrates</u>.

3. Relationships of the tunicates to the chordates is shown through the <u>adult</u> form.

4. All chordates <u>are</u> vertebrates.

5. All vertebrates <u>are</u> chordates.

6. <u>All</u> vertebrates have an endoskeleton made of bone.

7. Vertebrates never have more than <u>4</u> sets of paired appendages.

8. A cyclostome refers to a type of <u>organ</u>.

9. Lamprey eels are <u>scavengers</u>.

10. The shark's skeleton is made of <u>bone</u>.

11. The words placoid, dermal and ganoid describe types of <u>fins</u>.

12. The most highly organized of all the fish are the <u>bony</u> fish.

13. Most of the teleosts are <u>ovoviviparous</u>.

14. Metamorphosis of the tadpole is controlled by the <u>gastric</u> gland.

15. Sphenodon is the most <u>advanced</u> of the reptiles.

16. In mammals the allantois evolved into the <u>chorion</u>.

17. Fertilization in most fish takes place <u>internally</u>.

18. A cloaca is a (an) <u>tube</u>.

19. Egg white is the protein <u>gelatin</u>.

20. A group of eggs laid at one time by a bird is a <u>catch</u>.

21. A tadpole has a <u>three</u>-chambered heart.

22. Frog eggs are coated with a <u>cellulose</u>-like substance.

23. In mammals the <u>abdominal</u> cavity houses the lungs and heart.

24. Mammals are <u>warm-blooded</u> animals.

25. Only <u>reptiles</u> and mammals have a four-chambered heart.

Answers to Questions for Review

PART A

1. notochord
2. bilateral
3. vertebrae
4. vestigial
5. liver
6. jawless
7. scavengers
8. no
9. mud
10. placoid
11. lateral line
12. coelacanth
13. bony
14. tadpole
15. frogs and toads
16. amnion
17. reptiles
18. turtle
19. feathers
20. duckbill platypus
21. fry
22. oviducts
23. courtship
24. milt
25. placenta

PART B

1. c
2. a
3. c
4. d
5. b
6. a
7. b
8. d
9. c
10. d
11. b
12. c
13. b
14. b
15. d
16. c
17. d
18. d
19. b
20. c
21. a
22. a
23. a
24. c
25. b

PART C

1. brain
2. invertebrates
3. larval
4. are not
5. true
6. Most
7. two
8. fish
9. predators
10. cartilage
11. scales
12. true
13. oviparous
14. thyroid
15. primitive
16. placenta
17. externally
18. opening
19. albumen
20. clutch
21. two
22. jelly
23. thoracic
24. true
25. birds

CHAPTER 11

HOMO SAPIENS: A SPECIAL VERTEBRATE

Humans are *primates*. They share this mammalian order with gorillas, chimpanzees, monkeys, lemurs, tarsiers, slow lorises and the tailed bush babies. What are the special features that distinguish a primate from other mammals?

Scientists believe that at one time all primates lived in trees. The successful species were those that developed adaptations for arboreal life. Over evolutionary time, species such as chimpanzees, baboons and humans have left the trees and have adapted to life on land. However, certain characteristics persist in the ground-living species that show close relationship to their tree-living "cousins".

Fig. 11.1 illustrates the comparative shapes of hands of some primate species. These hands were adapted for grasping objects such as tree limbs, enabling the animals to locomote by swinging from tree limb to tree limb. Even the feet and tails of some primate species are *prehensile*, adapted for grasping. A second primate characteristic is the well-developed sense of sight. The eyes of most other mammals are located at the sides of the head. The eyes of a primate are directed forward, a structural arrangement that permits *stereoscopic vision*. This means that primates can see in three dimensions (length, width and breadth), a characteristic that enables them to see ahead with clarity the branches that they grasp. Another primate adaptation is the development of a larger and more *convoluted* brain. Convolutions are folds in the brain which increase surface area and allow for a greater number of nerve cells. Mention should also be made of two other primate characteristics. Primates have teeth that are less specialized for tearing and more useful for the grinding and chewing of a varied diet. Primates also have a small number of offspring and provide extended parental care for the young.

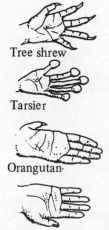

Tree shrew

Tarsier

Orangutan

Human

Fig. 11.1 A comparison of primate hands

A SPECIAL PRIMATE

Though humans are primates, they are set apart from other primates by some unusual and important adaptations. *Bipedalism*, the ability to walk on two legs (instead of four), has freed the forelimbs for doing work. The human can walk great distances and carry things from region to region—things that aid in the causes of hunting, or gathering or building.

Humans can live in the most forbidding of environments and are the most ecologically versatile of all of the primates. The human adjustment to various climates and land surfaces is probably due to the greater brain development. In all probability, brain development is the single greatest factor contributing to human's ability to form words and to speak. Speech is tied to a myriad of intricate brain functions: learning words, associating ideas, remembering. The formulation of the spoken word leads to the creation of the written word. When ideas are written, they are not lost in time. The human being is the only animal that can act in terms of history. Human civilizations function in terms of learned behavior called *culture*.

BASIC HUMAN STRUCTURE

The human body is divided into a *head*, *neck*, *trunk* and two pairs of appendages—namely, the *arms* and *legs*. Hair is present on the head, under the arms, sparsely on the arms and legs, and around the pubic area; in males, there is a greater distribution of body hair, often on the chest and heavily on the back of the hands, on the arms and the legs and on the face. The face is directed forward in a vertical position, the main axis of the body is set vertically, also.

In humans, as in other mammals (birds, too), the coelomic cavity is divided into three different areas. The *pericardial* cavity encloses the heart; *two pleural cavities* contain the lungs; and the *peritoneal* cavity holds the major part of the digestive system, the reproductive system and the urinary system.

The Skeletal System

The human skeleton, like that of all vertebrates, is a living endoskeleton that grows with the body. At birth, the human baby has a body that is made up of 270 bones. Due to the fusion of separate bones, the mature skeleton is composed of 206 bones. Fig. 11.2 shows some of the bones that make up the skeleton. Table 11.1 provides a more complete summary.

The human skeleton is a magnificent feat of engineering. The primary purpose of the skeleton is to carry the weight of the body and to support and protect the internal organs. The skeleton must be strong and able to absorb reasonable amounts of shock without fracturing. At the same time, the body framework must be flexible and light enough in weight to permit movement. Skeletal bones move in response to muscles that work like levers, allowing a variety of movements such as walking, running, hopping, sitting, bending, lifting and stooping.

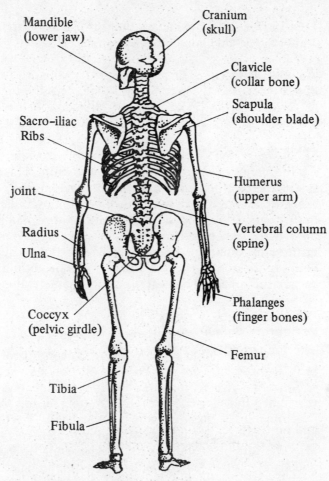

Fig. 11.2 Bones of the human body

TABLE 11.1.   The Human Skeleton

Skeleton	*Region*		*Names of Bones*
Axial Skeleton	Skull	Cranium	Frontal, parietal (2), temporal (2), occipital, sphenoid, ethmoid
		Face	Nasal, vomer, inferior turbinals, lacrimals, malars, palatines, maxillae, mandible, hyoid
	Vertebral Column		Vertebrae: cervical, thoracic, lumbar, sacral, coccyx
	Thorax		Ribs, sternum
Appendicular Skeleton	Pectoral Girdle		Clavicle, scapula
	Pectoral Appendages		Humerus, radius, ulna, carpals, metacarpals, phalanges
	Pelvic Girdle		Ilium, ischium, pubis (compose the os innominatum)
	Pelvic Appendages		Femur, patella, tibia, fibula, tarsals, metatarsals, phalanges

PARTS OF THE SKELETON

The human skeleton is divided into two major parts: the *axial skeleton* and the *appendicular skeleton*.

Axial Skeleton

The skull, the thorax (rib cage) and the vertebral, or spinal column are the three regions of the axial skeleton.

Skull

All the bones of the head compose the skull. The two regions of the skull—the *cranium* and the face—are made up of 22 flat and irregularly shaped bones. Eight bones form the cranium, which functions in the protection of the brain. The facial region, designed to protect the eyes, nose, mouth and ears, is composed of 14 bones. The *sinuses* are air spaces in the facial bones which aid in reducing the weight of the skull. The bones of the middle ear that function in transmitting sound to the inner ear are the smallest bones in the body—namely, the *hammer*, *anvil* and *stirrups*.

Vertebral Column

The vertebral column is composed of 26 bones known as vertebrae. At birth the vertebral column consists of 33 bones: seven cervical (neck) vertebrae, twelve thoracic vertebrae, five lumbar vertebrae, five sacral vertebrae, and four coccygeal (tail) vertebrae, but the five sacral bones fuse into one large triangular bone—the *sacrum*—at the back of the pelvis and the four coccygeal bones fuse into a single *coccyx*.

The vertebral column is the backbone made flexible by the cartilage and ligaments that join the individual vertebrae. Such a flexible backbone permits movement of the head and the bending of the trunk. Of major importance is the backbone's function in the protection of the spinal cord, which extends downward from the brain through the opening in each vertebra. Nerves branching from the spinal cord radiate to all parts of the body through openings in the sides of the vertebrae. Fig. 11.3 shows the discs of cartilage that separate the individual vertebrae. These discs prevent friction by the rubbing of the bones and serve as shock absorbers.

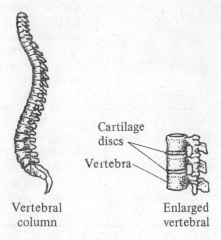

Cartilage discs

Vertebra

Vertebral column

Enlarged vertebral

Fig. 11.3 The human vertebral column

Thorax

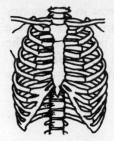

Fig. 11.4 Human rib cage

Just below the neck are 12 pairs of ribs that are attached to the vertebral column. The general shape of the *thoracic basket*, or rib cage, is shown in Fig. 11.4. The first ten pairs of ribs are attached to the breastbone, known also as the *sternum*, by cartilage strips, forming a structure that is smaller on top. The loose connections of the ribs to the vertebral column and the flexible cartilage connections at the sternum allow the ribs to move when the lungs are inflated. The 11th and 12th pair of ribs are often referred to as "floating ribs" because they are attached to the vertebral column but not to the breastbone.

The Appendicular Skeleton

The arms and hands, the legs and feet, and the bones of the shoulder and the pelvis make up the *appendicular skeleton*. You probably realize that "appendicular" is the adjective of the word *appendage*. An appendage is an attachment to a main body or structure. Legs and arms are attachments to the axial skeleton. The sites where arms and legs are attached to the axial skeleton are bones referred to as *girdles*. The *pectoral* (shoulder) girdle where the arms are attached is composed of the *scapula* (shoulder blade), a large triangular bone, and the *clavicle* (collarbone), a smaller curved bone. The legs are attached to the *pelvic girdle*, which is formed by the fusion of three bones: the ilium, the ischium, and the pubis on each side of the midline of the body (Fig. 11.5).

The bones of the legs and arms are appropriately called the *long bones*. The *femur*, the long bone between the hip and the knee is the longest and strongest bone in the body, supporting the weight of the body. The long bones in the lower leg are the thinner *fibula* and the thicker *tibia*.

Between the shoulder blade and the elbow is the bone of the upper arm, the *humerus*, a thinner version of the femur. The two long bones of the lower arm are the *radius* and the *ulna*. Finger and toe bones are known as *phalanges*; the bones of the foot, the *metatarsals*, while *metacarpals* are the bones of the hand.

In general, long bones are shaped like tubes with rounded processes at the ends which are designed to fit into other bones to form *joints*. The ends of the long bones are filled with *spongy bone*, a structural device which makes the bones light in weight but strong. The open spaces in spongy bone are filled with red marrow; while yellow marrow fills the shaft of the long bone.

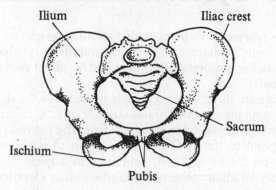

Fig. 11.5 The pelvic girdle

Bone marrow is an important substance. *Red marrow* in the spongy areas of the long bones and in the ribs and vertebrae are the sites where red blood cells are produced at the rates of millions per minute. The yellow marrow in the shaft of the long bones contains mostly fat. However, when the blood-making capability of the red marrow is low, yellow marrow is somehow converted into red marrow.

COMPOSITION OF BONE

About twenty percent of living bone is water; the remaining eighty percent consists of mineral matter and protein. Mineral matter deposited in bone usually forms the compound tricalcium phosphate $(Ca_3(PO_4)_2)$. Magnesium and other elements may also contribute to the mineral composition. The protein portion of the matrix is made of *collagen fibers* that are found in tendons, skin and connective tissue. The protein fibers and the mineral matter form the nonliving matrix of bone. However, bone also contains a variety of living cells and blood vessels that provide the pathways for nourishment to the cells and permit the removal of respiratory wastes from the cells.

Living bone cells are nestled in small spaces in the mineral matrix of bone. These cells are of three types, each adapted to carry out a specific function that has to do with the building and maintaining of bone. New bone material and the repair of broken bones is the work of *osteoblasts*, which are responsible for secreting the mineral and protein compounds that form the matrix. A second type of bone cell is the *osteoclast*, a bone breaker, able to dissolve bits of bone that are in the way of the efficient design of the skeleton. The destructive work of the osteoclasts is often followed by constructive work of the osteoblasts in the rebuilding of bone. The third type of bone cell, the *osteocyte*, functions as the caretaker of the bone tissue nearby.

The surface of nearly all parts of bone is covered by a tough membrane, the *periosteum*. The periosteum is perforated by microscopic blood vessels that supply the bone cells with nourishment. Although bone matrix appears to be solid, it is pierced by a network of *Haversian canals* through which blood vessels pass. Nerve fibers also extend into the bone interior through Haversian canals. The larger blood vessels pass directly into spongy bone and into the yellow marrow areas of the long bones.

The Muscular System

Muscles represent 40 percent of the total weight of the human body. Muscle tissue is characterized by contractility and electrical excitability, two distinctive properties that enable it to effect movement of the body and its parts.

There are three types of muscle tissue: smooth, striated, and cardiac. The movements of smooth and striated muscle tissue are controlled by contractile proteins and innervation from the nervous system; these muscles respond to the electrical stimulation of a nerve impulse. Cardiac muscle, on the other hand, functions to a great degree because of its own inherent ability to generate and conduct electrical impulses.

TYPES OF MUSCLE

Smooth Muscle

Smooth muscle is present in the walls of the internal organs, including the digestive tract, reproductive organs, bladder, arteries and veins. Because smooth muscle is contained in organs that do not respond to the will of a person, these muscles are called *involuntary muscles*. The most common function of smooth muscle is to squeeze, exerting pressure on the space inside the tube or organ it surrounds. Food is moved down the esophagus by the squeezing action of smooth muscle. The action of smooth muscles causes urine to be expelled from the bladder, semen to be discharged from the seminal vesicles and blood to be pumped through arteries. The opening and closing of the iris of the eye in response to light is accomplished by the action of smooth muscles.

Smooth muscle is made up of cells packed with contractile proteins, the cells forming sheets of tissue. Smooth muscle tissue is *innervated* by nerve cells and fibers from the sympathetic nervous system, that part of the nervous system that controls the activities of the internal organs. The contraction of smooth muscle is in response to stimulation by nerve cells, neurohumors or hormones.

Striated Muscle

Striated muscle is variously referred to as striped muscle, *voluntary muscle* or *skeletal muscle*—terms describing its structure and function. Located in the legs, arms, back and torso, striated muscles attach to and move the skeleton; since they are moved by the will of the person, they are often termed *voluntary muscles*. A striated or skeletal muscle is made up of a great number of *muscle fibers*, each of which extends the entire length of the muscle. There are probably around six billion fibers in more than 600 muscles scattered throughout the body.

If you look at a bit of striated muscle through the microscope, you will note that the muscle fibers contain many nuclei that seem not to be separated from each other by a plasma membrane. Such an arrangement in which plasma membranes are missing is called a *syncytium*. Each muscle fiber is innervated by at least one motor neuron. A *neuromuscular junction* is the space (synapse) between a nerve cell and a muscle.

Cardiac Muscle

Cardiac muscle is present only in the heart, where the cells form long rows of fibers. Unlike other muscle tissue, cardiac muscle contracts independently of nerve supply since reflex activity and electrical stimuli are contained within the cardiac muscle cells themselves. *Purkinje fibers*, part of the mechanism that controls heartbeat, are so specialized for conducting electrical impulses that they do not have contractile proteins. Each heartbeat is started by self-activating electrical activity of the heart's *pacemaker* known as the *sinoatrial node* (S-A node), positioned in the wall of the right atrium. From the S-A node, the impulse spreads throughout the atrium to the *atrioventricular node* (A-V node), a specialized bundle of cardiac muscle located on the atrium near the ventricles. The impulse spreads from the A-V node to all parts of the ventricles, causing simultaneous contractions in the ventricles.

Cardiac muscle is innervated by the tenth cranial nerve, the *vagus* nerve. The vagus is a mixed nerve containing some nerves that speed up heart beat and others that retard it. However, cardiac muscle contracts independently of nerve supply, and the effect of nerves on heartbeat is not too well understood.

HOW MUSCLES CONTRACT

The fine structure of a muscle fiber controls muscle contraction. A muscle fiber is made up of a bundle of finer fibers called *myofibrils*. A single myofibril is composed of smaller units named *sarcomeres* which are arranged in single file along the length of the myofibril. Within the sarcomere are alternating rows of thin and thick filaments. The thin filaments are attached to two vertical bands of thick protein called the Z *lines*. Fig. 11.6 shows the fine structure of a muscle myofibril.

Muscles contract due to a sliding filament mechanism. When the thick and thin filaments slide past each other, the Z lines of the sarcomeres are pulled closer together: in effect, contracting. When the sarcomeres contract, the myofibrils contract, which causes the contraction of muscle fibers. Fig. 11.7 illustrates the sliding filament mechanism of muscle contraction.

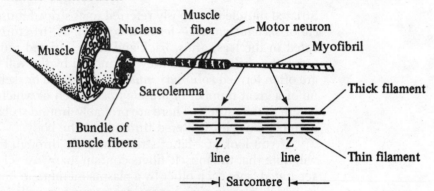

Fig. 11.6 The fine structure of skeletal muscle

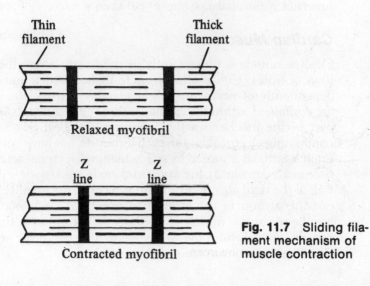

Relaxed myofibril

Contracted myofibril

Fig. 11.7 Sliding filament mechanism of muscle contraction

The Nervous System _____

The nervous system in humans is made up of two major parts: the central nervous system and the peripheral nervous system. Nervous tissue is specialized to receive stimuli from the outside environment and to conduct impulses to other body tissues. The development of the nervous system and particularly of the brain is what makes humans significantly different from other animals.

The basic unit of function of the nervous system is the *neuron*, or nerve cell. An understanding of the structure and function of this cell and how it transmits nerve impulses is important before dealing with the parts of the nervous system.

NERVE CELLS

The parts of the nerve cell are the *cyton*, or cell body; the *dendrites* and the *axon*. The dendrites receive signals from sense organs or from other nerve cells and transmit them to the cyton. The cell body passes signals to the axon, which then conducts the signals away from the dendrites and cell body. Notice that a nerve cell has only one axon and many dendrites. The axon terminating in *end brushes* (known also as *terminal branches*) is popularly called a *nerve fiber*. Many of the axons in the vertebrate body are covered by a fatty *myelin sheath* made of *Schwann cells*. The space between each Schwann cell is known as a *node of Ranvier*. Axons with the myelin sheath transmit impulses more quickly than those without the fatty coverings.

The nervous system has three types of neurons. *Sensory* or *afferent* neurons receive impulses from the sense organs and transmit them to the brain or spinal cord. *Associative* or *interneurons* are located within the brain or spinal cord. These transmit signals from sensory neurons and pass them along to motor neurons. *Motor* or *efferent* (Fig. 11.8) neurons conduct signals away from the brain or spinal cord to muscles or glands, so-called *effector* organs.

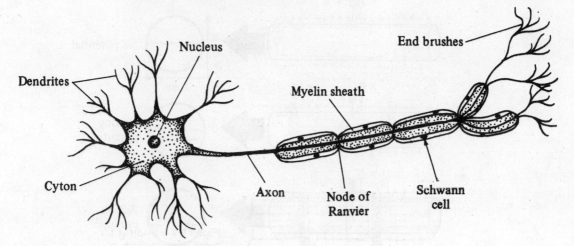

Fig. 11.8 Motor neurons

THE NATURE OF THE NERVE IMPULSE

The primary function of the nervous system is to permit communication between the external and internal environments. Communication in the nervous system is made possible by signals or impulses carried in a one-way direction along nerve cells. These impulses are electrical and chemical in nature.

When a neuron is not carrying an impulse, it is said to be at *resting potential*. When a nerve cell is stimulated to carry an impulse, its electrical charge changes and it is said to have an *action potential*. Action potentials (nerve impulses) from any one nerve cell are always the same. All impulses are of the same size, there being no graded responses. This is known as the "*all or none response*," meaning that a nerve cell will transmit an impulse totally or not at all (Fig. 11.9).

The electrical changes that occur in a nerve cell are due to differences in the distribution of certain ions, or charged particles on either side of the nerve membrane. Three conditions are responsible for the distribution of ions. First of these factors is the nature of the cell membrane itself. Remember, it is highly selective. The cell membrane is impermeable to the sodium ion (Na^+); at the same time, the cell membrane is highly permeable to the potassium ion (K^+). A second factor involves diffusion, the process by which molecules move from an area of greater concentration to one of lesser concentration. A third factor has to do with the attraction of ions of opposite charge and the repulsion of ions of like charge.

Let us begin with the situation in which there is a large concentration of K^+ inside of the cell and a large concentration of Na^+ on the outside of the nerve cell membrane. By diffusion, K^+ ions will cross the plasma membrane to the outside of the cell. The Na^+ cannot follow because the cell membrane is impermeable to them. K^+ will move to the outside until an equilibrium is established. This, in effect, causes the outside of the nerve cell membrane to be more positive than the inside of the membrane. The cell now can be described as being at its resting potential. The cell is also *polarized*.

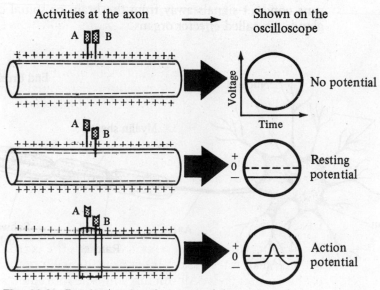

Fig. 11.9 Resting and action potentials

As an impulse travels through the nerve cell, several events take place. As the impulse touches a given point along the length of the plasma membrane, that site becomes permeable to Na^+ ions and Na^+ crosses the membrane and enters the cell. At the same time, the cell membrane becomes even more permeable to K^+ which then leaks out of the cell at a greater rate. These events lead to the *depolarization* of the cell in which there is a wave-like reversal of electrical charge along the length of the cell membrane. After the impulse is transmitted, the cell uses energy in mechanisms known as the sodium-potassium pump and carrier facilitated transport to bring the cell back to its resting potential.

Axons with myelinated sheaths can conduct impulses at the rate of 200 meters per second. Naked axons may conduct impulses at the rate of a few millimeters per second.

Impulses travel from one neuron to another crossing a specialized gap called the *synapse*. The synapse is a space between nerve cells that measures about 20 nanometers in width—just enough distance to prevent the touching of nerve cells. The terminal branches of the axons have synaptic knobs at their ends. As impulses travel along an axon, the synaptic knobs release chemicals called *neurotransmitters* or *neurohumors*. The neurotransmitters carry the impulse across the synapse onto the dendrites of the receiving nerve cell. The two main neurohumors are *acetylcholine* and *norepinephrine*, each with inhibitory or excitatory capabilities. After an impulse has crossed a synapse, the neurotransmitter is destroyed by an enzyme such as *cholinesterase*. (Many drugs of abuse, including LSD and cocaine, interfere with nerve impulse transmission.)

The endbrushes of the motor neurons are buried in a muscle or a gland. The junction between the nerve fibers and the muscle is known as the *neuromuscular junction*. The neurotransmitter *acetylcholine* carries the impulse from the nerve fiber to the muscle or gland. Nerve transmission across the neuromuscular junction is always excitatory, never inhibitory.

THE CENTRAL NERVOUS SYSTEM

The brain and the spinal cord compose the central nervous system. In the vertebrate body, the organs of the central nervous system are well protected by being wrapped in connective tissue and enclosed in bone. The brain, covered by the membranous *meninges*, rests in the skull cavity where it is enclosed by the cranium. The spinal cord, also covered by connective tissue, is circled by the vertebral column.

Brain

The human brain is divided into many parts, each with special functions. Among the most important parts are the cerebrum, cerebellum, and medulla (Fig. 11.10).

Cerebrum

In mammals the largest part of the brain is the cerebrum. The cerebrum is the seat of intelligence. It controls the voluntary, conscious activities of the human body. All voluntary muscle movements, all speech, thinking and memory are controlled by the cerebrum. It is also the center for

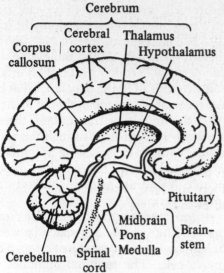

Fig. 11.10. The human brain

control of interpretation of all stimuli that enter the brain through the sense organs.

The cerebrum folds back on itself in many places forming wrinkles known as convolutions. These convolutions increase the surface area of the brain. The cerebrum is divided into two equal halves—the right and left cerebral hemispheres. Each of these hemispheres is subdivided into distinct regions of nervous control by fissures and convolutions. These regions control sensory areas for different parts of the body and are not haphazardly arranged.

The outer layer of the cerebrum is called the *cerebral cortex* and is composed of *gray matter*—unsheathed nerve cell bodies. Under the cortex is the *white matter* made up of sheathed axons. These fibers connect the various parts of the cerebrum and the cerebrum with other parts of the brain.

Cerebellum

The second largest part of the brain is the cerebellum. It is a bulbous mass of nerve tissue, which in human beings, lies underneath the back portion of the cerebrum. In general, the cerebellum controls balance. Its front and back areas regulate muscle tone. A region within the back lobe controls equilibrium; other sections coordinate voluntary movements. The cerebellum controls the precision and coordination of voluntary movements such as walking, running, dancing, skating, writing and typing.

Medulla

The medulla, or medulla oblongata, lies below the cerebrum and connects with the spinal cord. It contains a great number of *ganglia* (cytons) that receive sensory impulses and send out motor signals. Through the medulla pass many of the sensory and all of the motor nerves on their way to or from the higher centers in the brain. The medulla controls automatic, involuntary activities such as the contraction of smooth muscles, reflex movements, dilation and constriction of blood vessels, swallowing, breathing, and the like.

Other Parts

Other important parts of the brain include the thalamus and hypothalamus. The thalamus is the region of the brain where integration of sensory information occurs. The hypothalamus controls body temperature, osmoregulatory activities, maturity, thirst, hunger and sex drive. The hypothalamus is also the region where the nervous and hormonal systems interact.

Spinal Cord

The spinal cord is an elongated tubular structure containing masses of nerve cells and fibers and lying within the vertebral column. It is composed of a central H-shaped core of gray matter surrounded by white matter. The spinal cord conducts impulses to and from the brain; the impulses enter and leave the spinal cord through spinal nerves which extend from the spinal cord to the other organs of the body. The spinal cord is also the center for simple reflex activity.

In a simple *reflex*, often known as a *reflex arc*, only sensory nerves, the spinal cord, and motor nerves are involved. It allows instantaneous response without involving transmission to and from the brain. An example of a reflex is pulling your hand from a hot stove. When you touch a hot stove sensory nerves pick up the stimulus from receptors in the skin, transmit it to the spinal cord, which signals motor nerves to signal muscles for you to pull your hand away—all instantaneously.

THE PERIPHERAL NERVOUS SYSTEM

The peripheral nervous system connects the central nervous system—the brain and spinal cord—with the other organs of the body. It has two parts—somatic and automatic.

Somatic Peripheral System

The somatic peripheral system is composed of cranial nerves and spinal nerves. The fibers of both sensory and motor neurons are bundled together to form the cranial nerves. Twelve cranial nerves extend between the brain and the sense organs (eyes, ears, nose, etc.), heart, and other internal organs. Thirty-one pairs of mixed sensory and motor nerves extend from the spinal cord to the muscles and organs of the body. Each of the spinal nerves separates into sensory fibers and motor fibers as they join to form the spinal cord. The sensory fibers lead into the dorsal side of the spinal cord. Some of the sensory fibers synapse with the associative (interneuron) neurons; others lead to the brain. The cell bodies of the sensory nerves are in the dorsal root ganglia outside of the spinal cord. The motor nerves lead out from the spinal cord on the ventral side. Their cell bodies are in the spinal cord where they synapse with the associative neurons in the spinal cord.

Autonomic Nervous System

A network of nerves known as the autonomic nervous system controls the body's involuntary activities and the smooth muscles of the internal organs, glands and heart muscle. It is composed of motor (efferent) neurons leaving the brain and spinal cord and also of peripheral efferent neurons (Fig. 11.11).

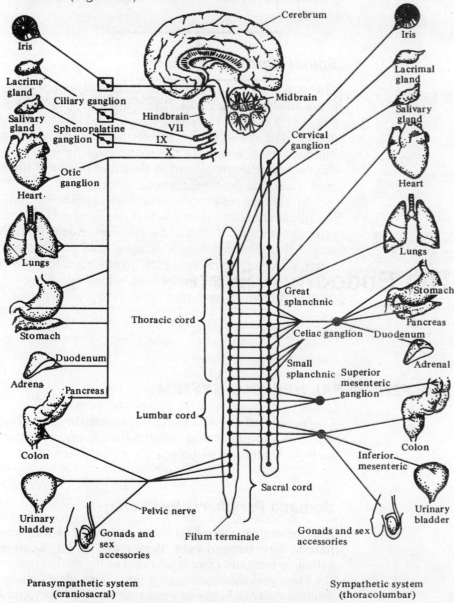

Parasympathetic system
(craniosacral)

Sympathetic system
(thoracolumbar)

Fig. 11.11 The autonomic nervous system

The autonomic nervous system is divided into the *sympathetic system* and the *parasympathetic system*. These subsystems are antagonists. When one set of nerves activates the smooth muscles of the body, the other set inhibits the action. For example: the parasympathetic nerves dilate the blood vessels and slow the heartbeat; the sympathetic nerves constrict the blood vessels and quicken heartbeat.

Fig. 11.12 summarizes the relationships of the parts of the nervous system.

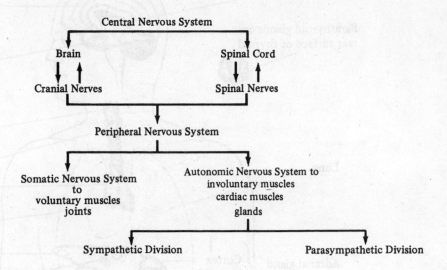

Fig. 11.12 The nervous system the easy way

The Endocrine System

Within the mammalian body there is a constellation of *ductless glands* known as the *endocrine system*. Fig. 11.13 shows the location of these glands in the human body. You will notice that these glands are not grouped together but are distributed throughout the body. Although these glands are not grouped together, they are considered to be a system because of similarities in structure and function.

As the name implies ductless glands do not have ducts and therefore do not discharge their secretions directly into another organ. This is in contrast to most of the glands in the body which are duct glands, delivering their secretions directly into a contiguous or nearby organ. For example, the salivary gland has a duct that delivers saliva directly into the mouth and sweat glands have ducts that conduct perspiration to the skin. Endocrine glands, also known as *glands of internal secretion*, deliver their secretions—hormones—into the bloodstream, which then carries them to their target organs. Hormones regulate many of the important metabolic activities of cells and organs.

The endocrine system is made up of the pituitary gland, thyroid gland, parathyroid glands, the adrenal gland, the isles of Langerhans in the pancreas, the thymus gland, the pineal gland, and the gonads—testes in the male and ovaries in the female. Certain secretions of the stomach and small intestine are also hormones and thus part of the endocrine system.

Through their secretions the endocrine glands regulate growth, rate of metabolism, response to stress, blood pressure, muscle contraction, digestion, immune responses, and the development and functioning of the reproductive system. Hormones exert their influence by becoming involved with the genetic machinery of cells and by affecting the metabolic activities of cells working through the cellular respiration pathways.

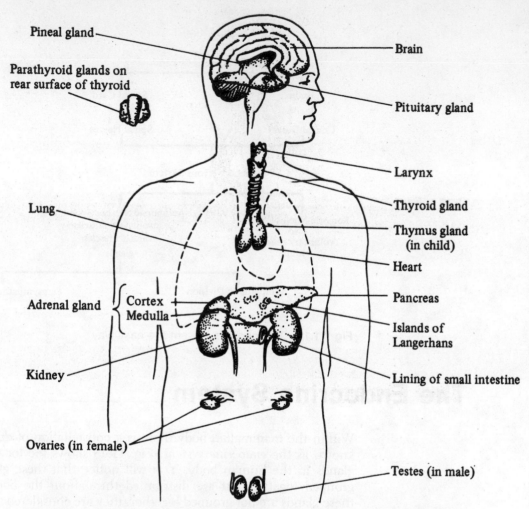

Fig. 11.13 Diagram showing the location of the major endocrine glands: pineal, pituitary, thyroid, parathyroid, thymus, pancreas (part), lining of the small intestine, adrenal glands, and sex glands

The pituitary gland, sometimes called the hypophysis, hangs from the base of the brain and is thought to exert control over much of the functioning of the other endocrine glands. It does this through its *trophic hormones*—hormones that stimulate the activity of other glands. Trophic hormones from the pituitary are known to stimulate secretions of the thyroid gland and of the adrenal gland and to regulate functioning of the sex organs. The pituitary, and in turn the rest of the endocrine system, is itself thought to be controlled by the hypothalamus region of the brain. The hypothalamus releases *neurosecretions*, known as *releasing factors*, which are transmitted to the pituitary where they, in turn, regulate the release of trophic hormones (Fig. 11.14).

The level of hormones in the blood and the release of hormones from endocrine glands is under a feedback control mechanism, with the blood "feeding" back to the brain information on how much hormone is circulating in the bloodstream. If, for example, the blood level of thyroxin, the thyroid hormone, falls too low, the hypothalamus secretes thyroid-releasing factor which stimulates the pituitary to release thyroid-stimulating hormone (TSH) and in turn the thyroid is stimulated to secrete more thyroxin. As the blood levels of thyroxin rise, the release of TSH and thyroxin is slowed.

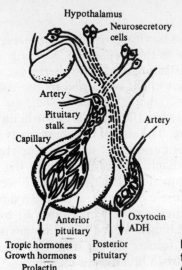

Fig. 11.14 The functional relationship between the pituitary and the hypothalamus

Table 11.2 summarizes the endocrine system, listing the major endocrine glands, their hormones and their functions, and the problems associated with excess or diminished secretion.

The Respiratory System

The process of respiration consists of external breathing and internal respiration. Breathing concerns the intake of air (*inhaling*) and the letting out of carbon dioxide and water vapor (*exhaling*). Internal respiration (cellular respiration) takes place in cells and is the series of biochemical events by which energy is released from food molecules.

Air is taken in through the nose. The nasal cavity is divided into two pathways by a septum made of cartilage and bone, and the bony *turbinates* increase the tissue surface along the dividing wall. The surfaces of the septum and the walls of the nasal cavities are covered with mucous membranes, some of which are lined with fine hairs. The nasal cavities have several small openings that lead to spaces in the facial bones called *sinuses*. These eight sinuses help to equalize the air pressure in the nasal cavity, reduce the weight of the skull, and contribute to the sound of the voice.

As air passes through the nasal cavity, it is warmed, humidified and filtered for dust particles. Incoming air passes through the nasal cavity into the *pharynx* (throat) through the *larynx* (voice box) and then into the *trachea* (windpipe). The trachea extends from the back of the throat down into the chest. It is held open by a series of C-shaped cartilage rings.

Just behind the middle of the breastbone, the trachea divides into two branches: the left and right *bronchi* (bronchus, sing.). Each bronchus divides and subdivides into smaller tubules called *bronchioles* that ramify throughout the lungs. Each bronchiole ends in a tiny air sac called an *alveolus*. Each alveolus is surrounded by blood capillaries. Oxygen diffuses from the lungs into the bloodstream and is transported to all parts of the body by way of the red blood cells. Conversely, carbon dioxide diffuses out of the blood into the air sacs and makes the reverse trip through the respiratory tubes, finally leaving the body through the nose.

The human body has two lungs. Each of these is enclosed in a double membranous sac known as the *pleural sac*. Not only is this sac airtight, but it also contains a lubricating fluid. The pleural sac and the lubricating fluid prevent friction that might be caused by the rubbing of the lungs against the chest wall.

The respiratory centers in the medulla of the brain and in other brain regions control breathing. The size of the chest cavity is regulated by the *diaphragm*, a flat sheet of muscle that separates the chest cavity from the abdominal cavity. The diaphragm is attached to the breastbone at the front, to the spinal column at the back and to the lower ribs on the sides. When the diaphragm muscle contracts, it is drawn downward creating a partial vacuum in the chest cavity. This causes air to flow through the respiratory tubes into the lungs. When the diaphragm is relaxed, the chest cavity becomes smaller, forcing the air out. The average rate of respiration in humans is about 18 breaths per minute.

TABLE 11.2.   Endocrines and Their Hormones

Name of Gland	Location	Hormone	Normal Function	Excess Secretion	Diminished Secretion
Anterior pituitary	Base of brain forward portion	Growth hormone (STH)	Affect skeletal growth, protein synthesis, blood glucose concentration	Gigantism acromegaly	Dwarfism
		Trophic hormones TSH ACTH FSH LH	Stimulate target glands thyroid adrenal cortex ovarian follicles; testes gonads	Oversecretion of glands	Undersecretion of glands
Posterior pituitary	Hind portion	Vasopressin	Control of blood pressure; reabsorption of water by kidney tubules	Increased blood pressure; glycogen converted to sugar	Decreased blood pressure; excess sugar changed to fat; kidney tubules not reabsorbing water
		Oxytocin	Contractions of uterus		
Thyroid	2 lobes on either side of larynx	Thyroxin (65% iodine)	Controls rate of oxidation in cells	Increased oxidation; nervous exophthalmic goiter	Lowered oxidation; in a child-cretinism; in an adult myxedemic goiter due to lack of iodine in drinking water
Parathyroid	Four glands above thyroid	Parathyroxin	Regulates amount of calcium in blood	Trembling due to lack of muscular control	Contraction of muscles (tetany); death.

TABLE 11.2. Endocrines and Their Hormones (*Continued*)

Name of Gland	Location	Hormone	Normal Function	Excess Secretion	Diminished Secretion
Stomach	Mucous lining (mucoos)	Gastrin	Stimulates secretion of gastric juice	Promotes ulceration of stomach wall	Inhibits gastric digestion
Small intestine	Mucous lining	Secretin	Activates the liver and pancreas to secrete and release their secretions	Excessive pancreatic and liver secretions	Diminished pancreatic and liver secretion
Adrenal medulla	Two glands above kidney	Adrenalin	Controls release of sugar from liver; contraction of arteries; clotting	Increases blood pressure; promotes clotting; releases glycogen; strengthens heart beat	
Adrenal Cortex		Glucocorticoids	Affects normal functioning of gonads; helps maintain normal blood sugar levels		Addison's disease: muscular weakness, darkening of skin, low blood pressure; death
		Mineralo-corticoids	Stimulates kidney tubules to reabsorb sodium		
Pancreas Isles of Langerhans	Embedded in pancreas	Insulin	Regulates storage of glycogen in liver; accelerates oxidation of sugar in cells		Diabetes; unused sugar remains in blood and is excreted with urine
Gonads	Abdominal region	Testosterone (Males) Estrogen Progesterone (Females)	Regulates normal growth and development of sex glands; regulates reproduction; controls sex characteristics	Premature development of gonads; effects on secondary sex characteristics	Interference with normal reproductive functions; diminished growth of sex characteristics
Thymus	Chest region	Thymosin	Stimulates immunological activity of lymphoid tissue		Breakdown of immune system
Pineal	Base of brain	Melatonin	Regulates gonadotropins by anterior pituitary		

The Circulatory System _____

The human circulatory system consists of the heart and the system of blood vessels that transport blood throughout the body.

THE HEART

The human heart lies in the chest cavity behind the breastbone and slightly to the left. The heart is a bundle of cardiac muscles specialized for rhythmic contractions and relaxations known as *heartbeat*. The rate of average heartbeat is 72 times per minute.

Fig. 11.15 shows the external structure of the heart. In size, the heart is about as large as a person's clenched fist. The walls are thicker on one side than on the other. The surface is covered with a number of small arteries and veins; these small arteries are the *coronary arteries* which carry blood laden with oxygen and nutrients to the muscle fibers of the heart. A number of large arteries and veins lead into the top of the heart. These carry blood to and from the other parts of the body.

The inside of the heart is divided into four chambers. The two chambers at the top are the receiving chambers, or the *atria*. The lower chambers, the *ventricles*, are pumping chambers. Each atrium is separated from the ventricle below by a valve. The atrium and the ventricle on the right are separated from the left atrium and ventricle by a thick wall of muscle called the *septum*.

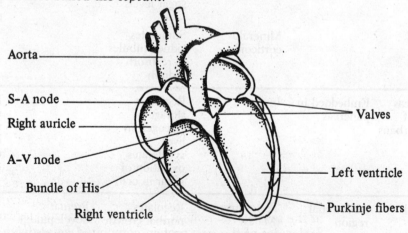

Aorta

S-A node

Right auricle

A-V node

Bundle of His

Right ventricle

Valves

Left ventricle

Purkinje fibers

Fig. 11.15 Structure of the heart

THE PATH OF THE BLOOD

The heart is a double pump. Blood flows from the right atrium into the right ventricle which pumps the blood through the pulmonary artery to the lungs where it receives oxygen and gives up waste products. The right ventricle represents the first pump. Blood returns from the lungs to the heart by way of the *pulmonary vein* and empties into the left atrium. The left atrium then sends the oxygenated blood into the left ventricle which then pumps it out to all parts of the body. The left ventricle represents the second pump.

Blood is oxygenated in the lungs. Deoxygenated blood leaves the heart (pumped by the right ventricle) by way of the pulmonary artery. Oxygenated blood is returned to the left atrium by way of the pulmonary vein. Arteries always carry blood away from the heart; veins carry it to the heart.

Heart valves prevent the backflow of blood. Separating the right atrium from the right ventricle is the *tricuspid valve*, so named because it has three flaps of tissue or cusps. The opening and closing of these cusps is controlled by papillary muscles. A valve with two cusps, the *bicuspid valve*, separates the left atrium from the left ventricle. (The bicuspid valve is also known as the *mitral* valve). Other valves are located where the aorta and the pulmonary arteries join the ventricles.

Several effects of the heart's pumping action can provide information about the condition of the heart and circulation. When a doctor listens to the heart through a stethoscope, he or she hears a series of sounds like "lubb-dub". The "lubb" is caused by the closing shut of the valves between the atria and the ventricles. At the same time the ventricles contract. The "dub" sound occurs when the semilunar valves of the aorta and the pulmonary artery close. The pause between the "dub" sound and the next "lubb" sound is the short time in which the heart rests.

The pulse is caused by the force of the blood on the arteries as the heart beats. The pulse rate, like that of heartbeat, is 72 beats per minute under normal conditions. Strenuous exercise increases the pulse rate.

Blood pressure measurement is a very valuable diagnostic procedure. Each time the ventricle contracts, blood is forced through an artery, increasing blood flow. The contraction phase is known as *systole*: relaxation, *diastole*. Normal systolic pressure is 120; diastolic, 80. This information is written as 120/80.

WORK OF THE BLOOD

Blood consists of a liquid medium called *plasma* and three kinds of blood cells: red blood cells (*erythrocytes*), white blood cells (*leucocytes*) and platelets.

The human body contains about 25 trillion erythrocytes, each one lasting about 120 days. New red cells are produced by the bone marrow at the rate of one million per second. Erythrocytes contain hundreds of molecules of the iron-protein compound *hemoglobin*. In the lungs, oxygen binds loosely to hemoglobin forming the compound oxyhemoglobin. As erythrocytes pass body cells with low oxygen content, oxygen is released from hemoglobin and diffuses into tissue cells. Carbon dioxide combines with another portion of the hemoglobin molecule and is transported to the lungs where it is exhaled.

For each 600 red blood cells there is one white blood cell, numbering in the billions in the blood. There are five types of white blood cells, functioning to protect the body against invading foreign proteins. The amoeboid-like *neutrophils* and *monocytes* behave as phagocytes, engulfing bacteria and other foreign proteins. *Eosinophils* detoxify histamine-like secretions. *Lymphocytes* participate in immune responses, and *basophils* produce anticoagulants.

BLOOD CLOTTING

Platelets are the smallest blood cells. When a capillary is cut, the platelets collect at the site of the injury. There they break into smaller fragments and initiate the complicated chemical process of blood clotting in which more than 15 factors, including thromboplastin, calcium (Ca), and fibrinogen, are involved in the formation of a clot containing blood cells in a fibrin meshwork (Fig. 11.16).

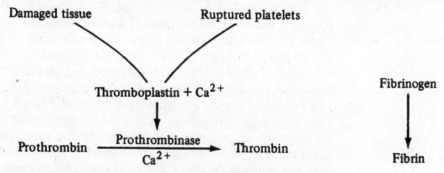

Fig. 11.16   Some steps of the blood clotting process

BLOOD TYPES

The main types of blood are A, B, AB, and O. Transfusions of blood are possible only when the blood types are compatible. If the blood types are not compatible proteins in the plasma will recognize foreign antigens on red blood cells and respond by causing the cells to *agglutinate*, or clump, a condition that causes blockage in small blood vessels. Table 11.3 summarizes the blood proteins involved in blood types.

The Lymphatic System _____

Homo sapiens actually has two circulatory systems. One is the blood circulatory system, the other is the lymphatic system. The body cells are bathed with tissue fluid called *lymph*. Lymph comes from the blood plasma, diffusing out of the capillaries into the tissue spaces in the body. Lymph differs from plasma in that it has 50 percent fewer proteins and does not contain red blood cells. Lymph has the important function of bringing nutrients and oxygen to cells and removing from them the waste products of respiration.

Although there is a constant flow of lymph from the blood plasma, neither the blood volume nor its protein content is diminished. This is so because as fast as lymph is drained from the blood plasma, it is returned to the blood. Radiating throughout the body are tiny lymph capillaries which join together to form larger lymph vessels and ducts. Tissue fluids are propelled through the body by differences in capillary pressure, muscle action, intestinal movements and respiratory movements. These movements squeeze the lymph vessels and push the fluid along. Lymph moves only in one direction toward the heart. Lymphatic valves prevent its backflow.

**TABLE 11.3. Proteins
of Blood Types**

Blood Type	Cell Antigen	Plasma Antibody
A	A	b
B	B	a
AB	AB	none
O	none	a and b

NOTE: Type AB—universal recipient
Type O—universal donor

Lymph is returned to the circulating blood through a large lymphatic vessel called the *thoracic duct*, which discharges its contents into the left subclavian vein. The right lymphatic duct empties its contents into the right subclavian vein. These veins then merge into other veins that empty into the heart.

In addition to vessels, the lymphatic system has lumpy masses of cells known as *lymph nodes* distributed throughout the body. These lymph nodes or glands are filtering organs that clear the tissue fluids of bacteria and other foreign particles. Lymph nodes are in the head, face, neck, thoracic region, armpits, groin and pelvic and abdominal regions. These nodes help the body in defense against disease. In addition to filtering out bacteria, they produce lymphocytes and antibodies.

Edema is the swelling that results from inadequate drainage of lymph from the body tissue spaces. Edema is brought about by heart and kidney disorders. malnutrition, injury, or other causes.

The Digestive System

The human digestive system begins with a mouth and ends with an anus, and is often described as a "tube within a tube". Variously called the *gut*, *alimentary canal* or the *gastrointestinal tract*, the digestive system extends from the lower part of the head region through the entire torso (Fig. 11.17).

Essentially, this system carries out five separate jobs that have to do with the processing and distribution of nutrients. First of all, it governs ingestion, or food intake. Secondly, it transports food to organs for temporary storage. Thirdly, it controls the mechanical breakdown of food and its chemical digestion. A fourth function is the absorption of nutrient molecules. Its final piece of work is the temporary storage and then elimination of waste products.

Digestion begins in the mouth. Teeth grind the food while three pairs of *salivary glands* pour salivary juice (saliva) into the mouth. Saliva contains the enzyme *salivary amylase* (ptyalin), which begins the digestion of starch. The moistened, chewed food is swallowed and moves through the throat into the food tube, or *esophagus*. The esophagus has no digestive function but moves the food into the stomach by waves of muscle contractions called *peristalsis*.

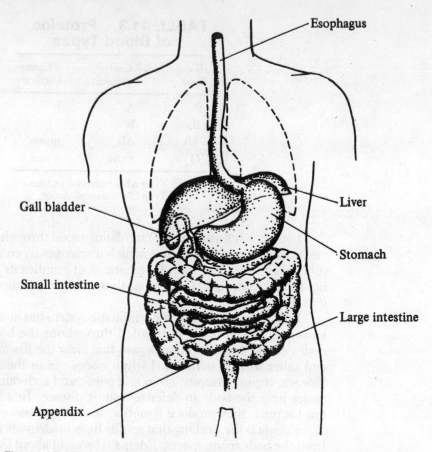

Fig. 11.17 The human digestive system

The *stomach* is the widest organ in the alimentary canal. It stores food while it churns and squeezes it, turning it into the consistency of a thick pea soup. In this semi-liquid form food can be worked upon by enzymes. *Gastric glands* embedded in the walls of the stomach secrete *gastric juice,* a combination of hydrochloric acid, and two enzymes, rennin and pepsin. Rennin is specialized for digesting the protein in milk, pepsin for hydrolyzing several plant and animal proteins. The semi-liquid food, often referred to as a *bolus* or *chyme,* is released a little at a time into the upper part of the small intestine.

Between the stomach and the small intestine is a ring of muscle called the *pyloric sphincter* that closes the stomach off from the *duodenum,* the upper part of the small intestine. In effect, the sphincter muscle regulates the flow of chyme from the stomach into the intestine.

The major work of digestion occurs in the small intestine. Lying outside of the alimentary canal are two important glands that are necessary for the many processes of digestion. The largest of these is the liver. It synthesizes *bile* and stores it in a pouch known as the *gall bladder.* Through bile ducts, bile is released into the small intestine where it serves as an emulsifier of fat, enabling it to be acted upon by the fat-digesting enzyme, lipase. The other accessory gland is the *pancreas,* a dual gland that synthesizes both hormones and enzymes. The pancreas releases pancreatic juice into the small intestine. This digestive juice is a combination of water and several digestive enzymes, each of which is specific for the digestion of fat, carbohydrate or protein.

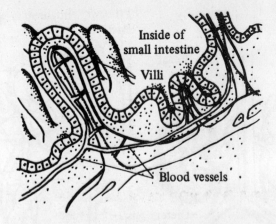

Fig. 11.18 A villus

In the walls of the small intestine are *intestinal glands* which manufacture and secrete intestinal juice, a combination of enzymes that digest starches, sugars and proteins. The outcome of all digestion is that nutrient molecules are reduced to soluble forms that enable them to cross cell membranes. Carbohydrates are digested into glucose or fructose. Proteins are broken down into amino acids. Fats are hydrolyzed into fatty acids and glycerol. These nutrients are absorbed by finger-like *villi* (villus, sing.), an adaptation in the small intestine for increasing surface area (Fig. 11.18).

Digested food diffuses into the capillaries of the villi. Blood then carries the food molecules to the liver through the *portal vein.* In the liver, sugar is removed from the blood and stored as *glycogen.* Digested fat molecules are absorbed into the *lacteals* (lymph vessels) and then enter the bloodstream through the *thoracic duct* which is in the chest cavity.

Food that is not digested passes into the large intestine, also called the *colon.* The large intestine absorbs a great deal of water and dissolved minerals. Undigested food or *feces* is pushed into the rectum where it is stored temporarily until eliminated through the anus.

The Excretory System

In human beings, the lungs, the skin and the urinary system work to expel the wastes produced in metabolic activities. The lungs excrete carbon dioxide and water. The skin expels water and salts from the sweat glands and a small amount of oil from the sebaceous glands. The urinary system handles the major work of excretion.

The human urinary system is located dorsally in the abdomen. Fig. 11.19 shows the organs that make up the urinary system and their locations. This system consists of two *kidneys,* tubes known as *ureters* extending from each kidney to a *urinary bladder* and a single *urethra,* a tube that leads out of the bladder.

The unit of structure and function of the kidney is the *nephron.* There are about one million of these microscopic units in each kidney. They actively remove waste products from the blood and return water, glucose, sodium ions and chloride ions to the blood. Fig. 11.20 shows the structure of a nephron.

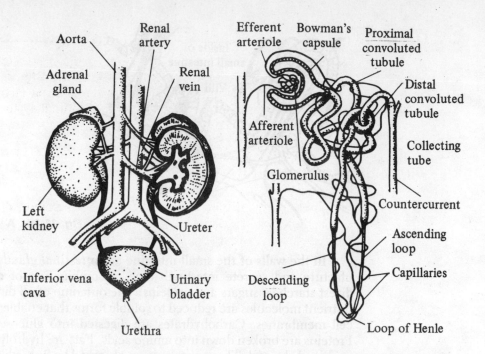

Fig. 11.19 The urinary system   **Fig. 11.20** The kidney nephron

The nephron is made up of several structures. The first of these is a knot of capillaries called the *glomerulus*. The glomerulus fits into a second portion—the *Bowman's capsule*, a cup-shaped cellular structure that leads into the third part, the kidney tubule. There are four main parts of each kidney tubule: the *proximal convoluted tubule*, the *loop of Henle*, the *distal convoluted tubule* and the *collecting duct*.

The *renal artery* transports blood into the kidney. The artery divides into smaller vessels called *arterioles* which divide into smaller blood vessels called capillaries. It is this network of capillaries that is called the glomerulus. The pressure of the blood in the glomerulus is quite high and forces fluid from the blood through the walls of the capillaries into the hollow cup of the Bowman's capsule. This process is known as *pressure filtration*. The fluid entering the nephron has the same composition of the blood except that it lacks blood cells, large protein molecules and lipids, which are not able to cross the membranes of the cells that compose the capillary walls.

As the filtrate (the water filtered out of the blood) moves into the proximal convoluted tubule, much of the water is reabsorbed back into the blood. Glucose and sodium are reabsorbed by active transport. As the water that is left in the tubule moves down into the *loop of Henle*, chloride ions and sodium are reabsorbed into the bloodstream by active transport. The remaining water and urea move into the distal convoluted tubule where additional sodium ions are reabsorbed into the bloodstream by active transport. The remaining water and urea waste pass into the collecting duct. From there it passes through the ureters to the bladder. This *urine* is now stored in the bladder until released from the body through the urethra.

The actions of the posterior pituitary hormone vasopressin and the steroid hormone aldosterone from the adrenal cortex regulate absorption in the kidney.

Sense Organs _____

The human body has five major senses—sight, hearing, taste, smell, and touch—that provide information about the external environment and transmit the stimuli to sensory nerves and ultimately to the brain for processing.

THE EYE AND VISION

The human *eyeball* measures about 2.5 centimeters in diameter. Most of the eyeball rests in the bony eyesocket of the skull. Only about one sixth of the eye is exposed. External structures associated with the eye are eyelids and lashes and eyebrows. A delicate protective membrane, the *conjunctiva*, covers the eye. Three pairs of small muscles attach the eye to the eyesocket. Secretions from tear glands help to keep the eye moist.

Fig. 11.21 shows the structure of the human eye. Notice that the eyeball is divided into two chambers that are separated by a *lens*. The front chamber contains a clear fluid called the *aqueous humor*. The back chamber contains a transparent jelly-like material called the *vitreous humor*.

The lens is transparent and is made of a great many layers of protein fibers. It measures about 8 millimeters in diameter. The function of the lens is to focus light on the *retina* at the back of the eyeball. The shape of the lens changes; it flattens when focusing on distant objects and thickens when focusing on near objects. The ability to bring objects into focus although they are located at different distances is called *accommodation*.

The colored portion of the front of the eye is the *iris;* in its center is a hole called the *pupil*. Light enters the eye through the pupil and passes through the cornea, the aqueous humor, the lens and the vitreous humor. Light reaches the *retina* where a barrage of signals are set up and are conducted to the *optic nerve* which carries these signals to the visual portions of the brain. The lens turns the image upside down and reverses it from left to right. The visual centers in the brain correct the inversions and reversals of the lens to make the image right side up.

Buried in the retina are cells called *rods* and *cones*. There are about 7 million cones and 120 million rods. The rods help the eye to accommodate in dim light and aid in night vision. The cones are responsible for color vision.

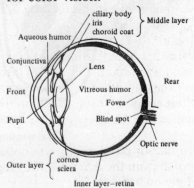

Aqueous humor

ciliary body
iris
choroid coat } Middle layer

Conjunctiva

Lens

Front

Vitreous humor

Rear

Fovea

Pupil

Blind spot

Optic nerve

Outer layer { cornea
sclera

Inner layer—retina

Fig. 11.21 Structure of the human eye

THE EAR AND HEARING

The human ear is made up of three divisions: the *outer ear*, the *middle ear* and the *inner ear*. Fig. 11.22 shows the structure of the ear. The outer ear catches sound waves and transports them to the *eardrum*, a membrane that stretches across the outer canal separating it from the middle ear. Sound waves cause the eardrum to vibrate.

The middle ear is a small cavity that is filled with air. It lies inside of the skull bone between the outer ear and the inner ear. At the bottom of the middle ear is an opening that leads into a canal. This canal is called the *Eustachian tube*, a passageway that connects the middle ear to the throat. This tube equalizes the air pressure in the ear with that of the throat. This is accomplished by yawning or swallowing.

The middle ear contains three very small bones called the *hammer*, the *anvil* and the *stirrup*. These are the smallest bones in the body. These bones accept the vibrations from the eardrum and transmit them to the oval window, one of two small membrane-covered openings between the middle ear and the inner ear.

The inner ear, which is entirely encased in bone, has a fluid-filled structure called the *cochlea*, so named because it resembles a snail in shape. The cochlea has numerous canals that are lined with hair cells. The vibrations from the oval window are transmitted to the hair cells in the cochlea and thence on to the *auditory nerve*, which conducts the vibrations to the brain. In the brain, these signals are interpreted into sounds.

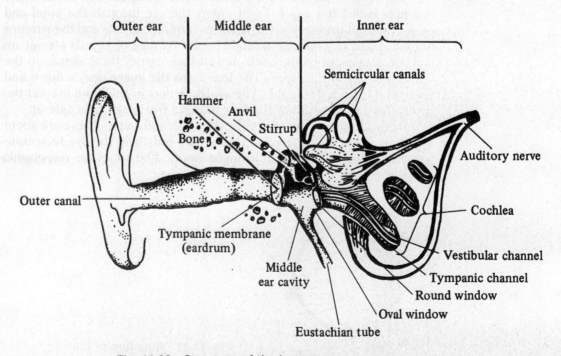

Fig. 11.22 Structure of the human ear

THE OTHER MAJOR SENSES

The organs of smell are located in the mucous membranes of the upper part of the nasal cavities. Special *olfactory cells* respond to odors and pass the impulse along the *olfactory nerve* to the brain. The sense of smell is far more important in lower animals than it is in humans.

The organs of taste are found chiefly on the tongue. These *taste buds*, as they are called, distinguish basic qualities such as bitter, sweet, sour, and salty.

The organs of touch are located on the skin surface and they respond to temperature, pain and pressure.

Reproduction

In this section we will discuss the reproductive system of humans and very briefly the development of the human young.

MALE REPRODUCTIVE SYSTEM

In the male reproductive system some organs are located outside of the body and others are positioned internally. Look at Fig. 11.23 and locate the scrotum. It is a sac-like organ that is located outside of the body. The *scrotum* contains the testes, glands that produce sperm and the male hormone *testosterone*. Also positioned outside of the body is the *penis*, the organ that delivers the sperm into the body of the female.

Each testis contains thousands of very small tubes called *seminiferous tubules*. Within these tubules, the sperm cells are manufactured. At the top of and behind each testis are coiled tubules called the *epididymis*. Each epididymis tubule functions as a storage place for sperm and also serves as a pathway which carries the sperm to a duct called a *vas deferens*. In its travels to the vas deferens, the sperm pass the *seminal vesicles* where they obtain nutrients. From the vas deferens, the sperm are conducted to the urethra, a single tube that extends from the bladder through the penis. Sperm cells leave the body through the penis. Erection of the penis and ejaculation of *semen* (a mixture of sperm and secretions from the seminal vesicle, the prostate gland and Cowpers gland) are processes necessary for the placement of sperm in the female reproductive tract.

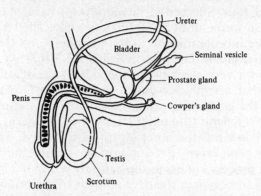

Fig. 11.23 Male reproductive system

FEMALE REPRODUCTIVE SYSTEM

The female reproductive system serves three important functions: the production of egg cells, the disintegration of nonfertilized egg cells, and the protection of the developing embryo. The reproductive system has specialized organs to carry out these functions.

Two oval-shaped ovaries lie one on each side of the midline of the body in the lower region of the abdomen. On a monthly alternating basis each ovary produces a mature egg. Eggs are located in spaces in the ovary called *follicles*. As an egg matures, it bursts out of the ovarian follicle and is released into the appropriate branch of the *fallopian tube*, a tube that leads from the region of the ovary to the uterus (Fig. 11.24).

If the egg is not fertilized, it is discharged from the body in a process called *menstruation*. The vascularized lining of the uterus, known as the *endometrium*, distintegrates in response to decreased levels of estrogen and progesterone in the blood. Menstrual bleeding lasts four to seven days.

If the egg is fertilized, it becomes implanted in the uterus where it goes through a series of cell divisions known as *cleavage*.

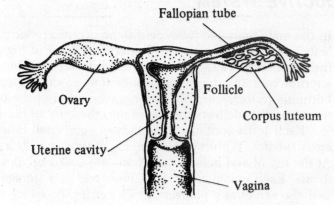

Fig. 11.24 Female reproductive system

CLEAVAGE

The fertilized egg goes through a series of cell divisions in which there is no growth in size of the zygote nor separation of the cells. Fig. 11.25 shows the stages of cleavage.

The first division of cleavage results in the formation of two cells; the second, four. Succeeding divisions result in eight cells, then 16, thirty two, and so forth, until a solid ball of cells called a *morula* is formed. Cells in the morula migrate to the periphery and the solid ball of cells changes to a hollow ball of cells called a *blastula*.

Within a very short time, one side of the blastula pushes inward, forming what resembles a double-walled cup. This stage of cleavage is known as the *gastrula*. During the gastrula stage three distinct layers of cells—the *ectoderm* (the outer layer), the mesoderm (the middle layer), and the endoderm (the inner layer)—are formed. These layers, known as *primary germ layers*, develop into the tissues and organs of the body through a process known as *differentiation*. An *embryo* is now forming. Table 11.4 summarizes the development of the primary germ layers.

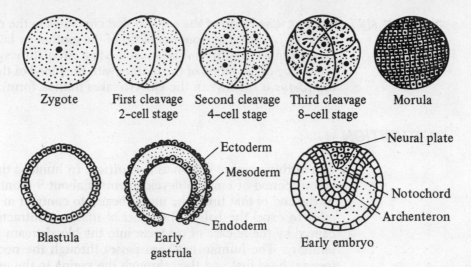

Zygote | First cleavage 2–cell stage | Second cleavage 4–cell stage | Third cleavage 8–cell stage | Morula

Blastula | Early gastrula | Early embryo

Ectoderm
Mesoderm
Endoderm

Neural plate
Notochord
Archenteron

Fig. 11.25 Stages of cleavage

TABLE 11.4.   Differentiation of the Three Primary Germ Layers

Ectoderm	Endoderm	Mesoderm
skin	lining of lungs	muscles
nervous system	lining of digestive system	skeleton
sense organs	pancreas	heart
	liver	blood vessels
	respiratory system	blood
		ovaries, testes
		kidneys

DEVELOPMENT OF THE HUMAN EMBRYO

As a result of cleavage of the fertilized egg and its implantation in the uterus many changes occur in the body. The follicle from which the egg cell burst becomes filled with some yellowish glandular material and is now known as the *corpus luteum*. The *corpus luteum* acts as an endocrine gland, secreting *progesterone* which prevents any other eggs in the ovary from developing further. The menstrual cycle is halted. No more eggs are discharged for the duration of the pregnancy.

The embryo produces several membranes that do not form any part of the new baby but which are necessary to the development and well being of the embryo. One of these membranes is the *amnion*, a water-filled sac that completely surrounds and protects the embryo. The water absorbs shocks and prevents friction that might damage the embryo.

The implanted embryo is attached to the uterus by means of the *umbilical cord,* a structure that contains blood vessels that function in carrying nutrients and oxygen to the embryo and transporting wastes away from the embryo. The umbilical cord connects with the *placenta,* a vascularized organ made up of tissues of the mother and tissues of the embryo. The blood of the embryo that circulates in the capillaries of the placenta is separated from the blood of the mother by layers of cells thin enough to allow diffusion between the two circulatory systems. There is no mixing of the blood of the mother with the blood of the embryo. (The term *fetus* is used when the embryo takes human form).

PARTURITION

The birth process is known as *parturition*. In humans the period of gestation (period of embryo development) is about 9 months or 40 weeks. At the end of that time, the uterus begins to contract in a process called *labor* to expel the baby. The onset of uterine contractions is probably caused by the release of oxytocin into the bloodstream by the posterior pituitary. The human newborn passes through the neck of the uterus (cervix) head first and then through the vagina to the outside.

MULTIPLE BIRTHS

Although humans usually produce only one offspring at a time, sometimes two, three, or even more young may be born at the same time. Of these multiple births, twins are the most common. There are two types of twins: identical and fraternal. Identical twins result from the fertilization of one egg and have the same genetic makeup. They are of the same sex and are almost identical in appearance. They develop in a common chorionic sac and share a common placenta. However, the umbilical cords are separate.

Fraternal twins develop from two separate fertilized eggs. They do not share a common genetic makeup and are no more alike than siblings born at separate times. The sexes may be different. Each fraternal twin has its own chorionic membrane and its own placenta.

Human Beings and Race

All humans belong to the species *Homo sapiens*. This means that the genetic material of all people is so similar that all humans can interbreed and produce fertile offspring. The human species is really a group of interbreeding populations. Populations that have adapted to certain environments become genetically different based on the frequency with which certain genes appear. Skin color, hair texture, body build and facial bone structure are a few of the characteristics that identify human population groups known as *races*.

Although we can make broad generalizations about the identifying characteristics of racial groups, not every member of each group fits these specifications. A set of physical characteristics can be drawn up that will fit individuals of several different races. Therefore, it is difficult for biologists and anthropologists to agree on the number of human races. Modern anthropologists divide *Homo sapiens* into three major stocks: Caucasoid, Mongoloid and African. Each of these groupings is subdivided into several human populations distinguishable by certain pronounced characteristics. The Caucasoid stock is composed of four white races: Nordic, Alpine, Mediterranean, and Hindu. The Asian stock is divided into the Malaysian, American Indian, and the Mongolian populations. The African stock is separated into Black, Melanesian, Pygmy Black, and Bushman populations. Of doubtful grouping are the Polynesian and the Australoid peoples. Many anthropologists classify these groups as Asian; others disagree. No matter the grouping, the differences among human races are very small.

Chronology of Famous Names in Biology

1637	**William Harvey** (English)—discovered how blood circulates in the body.
1759	**Philibert Gueneau de Montbeillard** (French)—made the first systematic measurements of the growth of a child.
1771	**John Hunter** (English)—dissected the body of the Irish giant, Charles Byrne, and discovered a much enlarged pituitary gland.
1790	**Luigi Galvani** (Italian)—discovered that electrical stimuli can cause muscles to contract.
1830	**Thomas Addison** (English)—discovered the effect of damaged adrenals on health, causing mottled skin, weight loss, anorexia, irritability.
1830	**Jan Evangelista Purkinje** (Czech)—discovered the sweat glands in the skin and the fiber network in cardiac muscle; demonstrated the importance of fingerprints.
1849	**Arnold Adolph Gerthold** (Germany)—discovered the endocrine function of testes.
1853	**Thomas Curling** (English)—first reported myxedema.
1855	**Claude Bernard** (French)—discovered the glycogenic function of the liver.
1860	**Anders Retzius** (Swedish)—devised and named the cephalic index: maximum length to maximum breadth of the skull.
1885	**Paul Langerhans** (German)—discovered the islets in the pancreas.
1886	**Pierre Marie** (French)—discovered that an oversecretion of growth hormone causes acromegaly.
1873	**Camillo Golgi** (Italian)—devised a method of impregnating metallic salts into nerve cells to better determine their structure. He discovered the dendrites and axons of nerve cells.
1882	**Richard Owen** (English)—discovered the existence of parathyroid glands by dissecting an Indian rhinoceros.

1904	**Ernest Starling** and **William Bayliss** (English)—devised the term *hormone*.
1914	**Edward C. Kendall** (American)—isolated thyroxin from cattle thyroid.
1915	**David Marine** (American)—discovered the link between iodine and goiter.
1921	**Frederick Barting** and **Charles Best** (Candian)—isolated insulin.
1925	**David Marine** (American)—associated iodine deficiency with the high incidence of goiter in Cleveland.
1929	**Edward Doisy** (American)—isolated the ovarian hormone estrone.
1930	**Karl Landsteiner** (American)—discovered human blood groups.
1942	**Charles Drew** (American)—devised a more effective method for preserving blood for transfusions.
1944	**Herbert Evans** (American)—discovered growth hormone.
1944	**Choh Hao Li** (American)—isolated growth hormone and other pituitary secretions.
1948	**Otto Loewi** (French)—discovered the presence and actions of neurohumors.
1950	**Andrew F. Huxley** and **R. Niedegarde** (English)—found that two lines of skeletal muscle fibers move close together when a muscle contracts.
1950	**Gregory Pincus** (American)—developed a steroid to suppress ovulation from the roots of a wild Mexican yam.
1955	**Alan L. Hodgkin** and **Andrew F. Huxley** (English)—won Nobel Prize for showing how a difference of potential contributes to the functioning of a neuron. They studied the giant axons of the squid.
1958	**Hans Selye** (Canadian)—discovered that stress affects the endocrine system.
1959	**Morris Goodman** (American)—prepared animal serums with antibodies which measure the degree of relationship among the various species of primates.
1965	**Hugh E. Huxley** and **Jean Hanson** (English)—developed the sliding filament theory of muscle contraction.

Words for Study

accommodation	epididymis	osteoclast
action potential	erythrocyte	osteocyte
afferent neuron	Eustachian tube	oval window
agglutinate	femur	parturition
alimentary canal	fetus	pericardial
alveolus	fibula	periosteum
antibody	follicle	peritoneal
appendage	gastrula	phalanges
atrium	girdle	platelet
axon	glomerulus	pleural
basophil	gray matter	prehensile
biscuspid valve	Haversian canals	primate
bipedalism	hormone	Purkinje fibers
blastula	humerus	pyloric sphincter
bronchus	hypophysis	race
cardiac muscle	hypothalamus	radius
cerebellum	interneuron	reflex
cerebrum	involuntary muscle	releasing factor
chyme	lacteals	retina
clavicle	lens	rods
cleavage	leucocyte	round window
cochlea	lymph	sacrum
coccyx	lymphocyte	sarcomere
cones	medulla	scapula
convolution	meninges	Schwann cell
coronary artery	menstruation	scrotum
corpus luteum	metacarpal	semen
cranium	metatarsal	sinoatrial node
culture	mitral valve	sinus
cyton	monocyte	skull
dendrite	morula	smooth muscle
diaphragm	myofibril	sodium-potassium pump
diastole	neuromuscular junction	stereoscopic vision
differentiation	neuron	sternum
duodenum	neurotransmitter	striated muscle
eardrum	neutrophil	synapse
edema	node of Ranvier	syncytium
effector organ	olfactory nerve	systole
endocrine gland	optic nerve	thalamus
eosinophil	osteoblast	thoracic duct

thorax	ureter	vertebral column
tibia	urethra	villus
tricuspid valve	vagus nerve	voluntary muscle
trophic hormone	vas deferens	white matter
ulna	ventricle	
umbilical cord	vertebrae	

Questions for Review

PART A.   **Completion.** Write in the word that correctly completes each statement.

1. Humans belong to the species named ..1..
2. The hands of primates are adapted for ..2..
3. The ability to walk on two legs is known as ..3..
4. Human patterns of learned behavior are known collectively as ..4..
5. The human pericardial cavity encloses the ..5..
6. One function of the facial sinuses is to reduce the ..6.. of the skull.
7. A common name for the thoracic basket is the ..7..
8. Bones that attach arms and legs to the axial skeleton are known as ..8..
9. Cells that build bone matrix are called ..9..
10. Cells that dissolve the mineral matrix of bone are the ..10..
11. The function of smooth muscle is to ..11.. organs it surrounds.
12. The type of muscle that moves the skeleton is called voluntary or ..12.. muscle.
13. The space between a nerve cell and a muscle is the ..13.. junction.
14. Two distinctive properties of muscle are electrical excitability and ..14..
15. The S-A node acts as the ..15.. of the heart.
16. Myofibrils are correctly associated with ..16.. cells.
17. Sensory information is integrated in that brain region known as the ..17..
18. The nature of a nerve impulse is both electrical and ..18..
19. Balance and coordination are controlled by the part of the brain known as the ..19..
20. The peripheral nervous system is divided into the somatic and ..20.. systems.
21. A distinctive structural characteristic of endocrine glands is that they do not have ..21..
22. The homeostatic mechanism that regulates the pituitary gland and its target organs is known as a ..22.. system.
23. The endocrine glands that cap the kidneys are the ..23..

24. The contraction of a ventricle in the heart is known as ..24..
25. The force of the blood on the arteries is known as ..25..
26. The largest lymphatic in the body is the ..26..
27. The unit of structure and function in the kidney is the ..27..
28. The part of the eye that bends light rays is the ..28..
29. The ..29.. in the eye are responsible for color vision.
30. Pressure in the middle ear is regulated by the ..30.. tube.
31. The solid ball of cells formed during cleavage is known as the ..31..
32. The corpus luteum behaves as a ..32.. gland.
33. The birth process is known as ..33..
34. Food moves through the digestive tract by waves of muscle contraction known as ..34..
35. The ..35.. closes the stomach from the duodenum.
36. Bile is stored in the ..36..
37. The large intestine is also called the ..37..
38. Tubes leading from the kidney to the urinary bladder are called ..38..
39. The ..39.. nerve carries signals from the eye to the brain.
40. Semen is a mixture of sperm and secretions from the seminal vesicle, Cowpers gland and the ..40.. gland.

PART B. **Multiple Choice.** Circle the *letter* of the item that correctly completes each statement.

1. The eutheria are
   (a) flying birds
   (b) infertile frogs
   (c) female kangaroos
   (d) placental mammals

2. A prehensile tail is one that is adapted for
   (a) swimming
   (b) grasping
   (c) gliding
   (d) balancing

3. Convolutions are best associated with the
   (a) pharynx
   (b) lung
   (c) brain
   (d) nerve cord

4. The number of bones in the mature human skeleton is
   (a) 206
   (b) 270
   (c) 290
   (d) 305

5. The major function of the backbone is to
   (a) protect the spinal cord
   (b) bend the trunk
   (c) to hold the neck and head
   (d) to support the legs

6. The strongest bone in the body is the
   (a) clavicle
   (b) sternum
   (c) femur
   (d) metatarsal

7. Which is a true statement about bone?
   (a) The matrix is composed of living material.
   (b) Yellow marrow is in the spongy areas of the long bones.
   (c) Bone has a variety of living cells.
   (d) The ends of long bones are made of solid bone.

8. Bone is covered by a membrane known as the
   - (a) peritoneum
   - (b) pericardium
   - (c) perithecium
   - (d) periosteum

9. Which of the following are three names for voluntary muscle:
   - (a) striped, cardiac, smooth
   - (b) striated, skeletal, smooth
   - (c) skeletal, cardiac, striated
   - (d) striated, striped, skeletal

10. A syncytium refers to
    - (a) a group of cells without plasma membrane boundaries
    - (b) a segment of smooth muscle that rings the trachea
    - (c) a group of nerve cells that stimulates muscle
    - (d) a single muscle fiber when stimulated by a neuron

11. Which of the following is a true statement about Purkinje fibers?
    - (a) They are found in smooth muscle.
    - (b) They lack contractile proteins.
    - (c) They form a syncytium with skeletal muscle.
    - (d) They cannot conduct electrical impulses.

12. If nerves to the heart are severed, the heart will
    - (a) cease functioning
    - (b) beat faster
    - (c) beat regularly
    - (d) beat erratically

13. Impulses enter nerve cells by way of the
    - (a) dendrites
    - (b) cytons
    - (c) axons
    - (d) terminal branches

14. Myelin is a nerve cell covering composed of
    - (a) $CaCO_3$
    - (b) collagen
    - (c) fat
    - (d) carbohydrate

15. The fact that a nerve cell transmits an impulse totally or not at all is a type of response known as the
    - (a) take it or leave it
    - (b) all or none
    - (c) make it and take it
    - (d) all or some

16. The width of a synapse is approximately
    - (a) 80 nm
    - (b) 60 nm
    - (c) 40 nm
    - (d) 20 nm

17. The part of the brain that is necessary for life is the
    - (a) cerebrum
    - (b) cerebellum
    - (c) fissure of Rolando
    - (d) medulla

18. The sympathetic and parasympathetic systems function
    - (a) antagonistically to each other
    - (b) to reinforce the activities of each other
    - (c) without relationship to the autonomic system
    - (d) to counteract the control of the cerebrum

19. The hypothalamus is a part of the brain where
    - (a) integration of sensory information occurs
    - (b) integration of memory and higher thought processes occur
    - (c) follicle stimulating hormone is released into the blood stream
    - (d) the nervous and hormonal systems interact

20. Blood in the pulmonary artery
    (a) lacks oxygen
    (b) lacks carbon dioxide
    (c) contains nitrogen
    (d) contains all three gases

21. The most important function of erythrocytes is to
    (a) carry nutrients from cell to cell
    (b) protect the body against disease
    (c) carry oxygen to all cells
    (d) remove carbon from all cells

22. The lymph nodes are glands that
    (a) secrete hormones and neurohumors
    (b) propel tissue fluids through the body
    (c) control the production of red blood cells
    (d) filter bacteria from the tissue fluids

23. Lymphocytes are white blood cells that
    (a) phagocytize bacteria
    (b) detoxify histamines
    (c) produce anticoagulants
    (d) participate in immune responses

24. Breathing is controlled by the
    (a) diaphragm
    (b) respiratory centers in the brain
    (c) level of carbon dioxide in the blood
    (d) all three of the above

25. The lungs are enclosed in a set of double membranes known as the
    (a) pericardium
    (b) periosteum
    (c) pleural sac
    (d) peritoneum

26. In human beings the organs of excretion are the
    (a) kidneys, lungs, rectum
    (b) large intestine, sweat glands, lungs
    (c) kidneys, lungs, sweat glands
    (d) sweat glands, kidneys, anus

27. The function of the urinary bladder is to
    (a) store urine
    (b) detoxify urea
    (c) add CO_2 to ammonia
    (d) filter out glucose

28. Blood is transported to the kidney from the dorsal aorta by the
    (a) renal vein
    (b) renal artery
    (c) arterioles
    (d) glomerulus

29. The cup-shaped portion of the nephron is the
    (a) loop of Henle
    (b) glomerulus
    (c) Bowman's capsule
    (d) convoluted tubules

30. The widest organ in the alimentary canal is the
    (a) stomach
    (b) large intestine
    (c) colon
    (d) gall bladder

31. Two glands lying outside the alimentary canal but important to digestion are the
    (a) liver and kidney
    (b) pancreas and thoracic duct
    (c) liver and pancreas
    (d) liver and colon

32. The major work of digestion occurs in the
(a) stomach
(b) small intestine
(c) large intestine
(d) esophagus

33. Digestion begins in the
(a) stomach
(b) small intestine
(c) esophagus
(d) mouth

34. In the human male, sperm is stored in tubules known as
(a) sperm ducts
(b) Cowper's gland
(c) vas deferens
(d) epididymis

35. Follicles are sites where
(a) fertilization occurs
(b) chromosomes are halved
(c) ova are produced
(d) sperm receive nutrients

36. The primary germ layer that gives rise to the blood vessels is the
(a) endoderm
(b) mesoderm
(c) ectoderm
(d) protoderm

37. The membranous sac of water that surrounds the developing human fetus is the
(a) amnion
(b) chorion
(c) allantois
(d) placenta

38. Identical twins are produced from
(a) two eggs fertilized by two sperm
(b) two eggs fertilized by one sperm
(c) one egg fertilized by one sperm
(d) one egg fertilized by two sperm

39. The implanted embryo is attached to the uterus by means of the
(a) placenta
(b) umbilical cord
(c) yolk stalk
(d) allantois

40. Modern anthropologists divide *Homo sapiens* into three major stocks:
(a) Caucasoid, Mongoloid, African
(b) Caucasoid, Bushman, Melanesian
(c) Caucasoid, Hindu, Mongolian
(d) Caucasoid, Nordic, Pygmy Black

PART C. Modified True-False. If a statement is true, write "true" for your answer. If a statement is incorrect, change the *underlined* word to one that will make the statement true.

1. Scientists believe that at one time all primates lived in savannas.

2. Due to stereoscopic vision, primates can see in two dimensions.

3. The primate with the best developed brain is the chimpanzee.

4. The hammer, anvil, and stirrups are sinuses in the middle ear.

5. The humerus is the long bone in the lower leg.

6. Red blood cells are produced in the red marrow of long bones.

7. "Bone breakers" are cells known as osteoblasts.

8. Striated muscle is called involuntary muscle.

9. The sympathetic nervous system innervates <u>striated</u> muscle.

10. The type of muscle that makes up the heart is <u>smooth</u> muscle.

11. Z lines are bands of <u>fat</u>.

12. The unit of function and structure of the nervous system is the <u>cyton</u>.

13. Schwann cells are best associated with the <u>myofibril</u> sheath.

14. Nervous impulses are conducted by <u>efferent</u> neurons to the brain or spinal cord.

15. Motor neurons conduct impulses to the <u>effectors</u>.

16. Interneurons are located <u>outside</u> of the brain and spinal cord.

17. The reversal of polarity on a nerve cell is known as the <u>resting</u> potential.

18. The brain and the <u>vertebral column</u> make up the central nervous system.

19. The membranes that cover the brain are known as the <u>phalanges</u>.

20. The parasympathetic and sympathetic systems are divisions of the <u>somatic</u> nervous system.

21. Glands of internal secretion release chemicals collectively known as <u>enzymes</u>.

22. Chemicals known as releasing factors are secreted by the <u>anterior pituitary</u>.

23. The presence of estrogen in the blood signals the <u>uterus</u> to secrete luteinizing hormone.

24. The coronary arteries are positioned <u>inside</u> of the heart.

25. The liquid part of the blood is called <u>serum</u>.

26. The life of a red blood cell lasts for <u>30</u> days.

27. White blood cells function to protect the body against invasion by foreign <u>worms</u>.

28. Platelets initiate the <u>clumping</u> of blood.

29. Another name for the voice box is the <u>pharynx</u>.

30. Each bronchiole ends in an air sac known as an ..30..

31. Rennin digests protein in <u>milk</u>.

32. The <u>gall bladder</u> synthesizes bile.

33. The gall bladder <u>stores</u> bile.

34. Digested fat molecules are absorbed into <u>villi</u> and then enter the bloodstream through the thoracic duct.

35. Vibrations are carried to the brain by the <u>optic</u> nerve.

36. The back chamber of the eye is filled with <u>aqueous</u> humor.

37. The three small bones in the middle ear are the anvil, <u>club</u>, and stirrup.

38. Each <u>identical</u> twin has its own chorionic membrane and its own placenta.

39. The human female has <u>one</u> ovary (ovaries).

40. In the male reproductive system some organs are located outside the body and some are positioned internally.

Answers to Questions for Review

PART A

1. *Homo sapiens*
2. grasping
3. bipedalism
4. culture
5. heart
6. weight
7. rib cage
8. girdles
9. osteoblasts
10. osteoclasts
11. squeeze
12. striated or striped
13. neuromuscular
14. contractility
15. pacemaker
16. muscle
17. thalamus
18. chemical
19. cerebellum
20. autonomic
21. ducts
22. feedback
23. adrenals
24. systole
25. pulse
26. thoracic duct
27. nephron
28. lens
29. cones
30. Eustachian
31. morula
32. endocrine
33. parturition
34. peristalsis
35. pyloric sphincter
36. gall bladder
37. colon
38. ureters
39. optic
40. prostate

PART B

1. d	15. b	28. b
2. b	16. d	29. c
3. c	17. d	30. a
4. a	18. a	31. c
5. a	19. d	32. b
6. c	20. a	33. d
7. c	21. c	34. d
8. d	22. d	35. c
9. d	23. d	36. b
10. a	24. d	37. a
11. b	25. c	38. c
12. c	26. c	39. b
13. a	27. a	40. a
14. c		

PART C

1. trees
2. three
3. human
4. bones
5. upper arm
6. true
7. osteoclasts
8. voluntary
9. smooth
10. cardiac
11. protein
12. neuron
13. myelin
14. afferent
15. true
16. inside
17. action
18. spinal cord
19. meninges
20. autonomic
21. hormones
22. hypothalamus
23. anterior pituitary
24. outside
25. plasma
26. 120
27. proteins
28. clotting
29. larynx
30. alveolus
31. true
32. liver
33. true
34. lacteals
35. auditory
36. vitreous
37. hammer
38. fraternal
39. two
40. true

CHAPTER 12

NUTRITION

*N*utrition is the totality of methods by which an organism satisfies the energy, fuel and regulatory needs of its body cells. Those substances that contribute to the nutritional needs of cells are the *nutrients*. Animals take these nutrients into the body by the *ingestion* of food. Food, therefore, refers to edible materials that supply the body nutrients. Nutrients needed in large amounts are classified as *macronutrients*: carbohydrates, proteins and fats. *Micronutrients*—vitamins and minerals—are needed in smaller amounts. Vitamins are organic compounds; minerals are inorganic. *Malnutrition* results from the improper intake of nutrients. This may be due to a person's eating too little food or to the intake of too much of one nutrient and not enough of others.

Nutritionists urge the eating of a *balanced diet*. A balanced diet is a good mixed diet that includes choices from the four major groups of food: the milk group, the meat group, the vegetable and fruit group and the breads and cereals group. Three or four servings from each of these groups each day will ensure a nutritionally useful diet.

Macronutrients

CARBOHYDRATES

Carbohydrates include starches and sugars. The primary function of carbohydrates in the diet is to serve as fuel for the body cells. All carbohydrates must be broken down into glucose or fructose by digestion before they can be used by cells. If body cells receive more simple sugar than they can use as energy, some of the excess sugar is stored in the liver and muscles as *glycogen*, commonly called *animal starch*. However, if carbohydrates are taken into the body in much larger quantities than the body needs, they are converted into fat and stored under the skin and around body organs.

Carbohydrates help form the structures of some important biological compounds, including parts of the cell membrane, and they assist the body in the manufacture of biotin and other of the B-complex vitamins.

Food sources of carbohydrates include potatoes, fruits, vegetables, cereal grains, beans, peas, sugar cane, beets, milk, baked goods and pasta.

PROTEINS

Proteins are the most abundant of the organic compounds in body cells. They compose all of the fibrous structures in the body including hair, nails, ligaments, the microfilaments in cells and the myofibrils of muscles. They also form part of hemoglobin, certain hormones such as insulin and thousands of enzymes that control biochemical processes of cells. Proteins are assimilated into protoplasm and are vital to the formation of DNA molecules. Proteins are also necessary in forming antibodies, molecules that constitute an important part of the immune system which functions to ward off disease, and in regulating the water balance and acid-base balance in the body.

Food sources of protein include meat, fish, eggs, milk, cheese, and soy, lima, and pea beans. A serious protein deficiency disease is *kwashiorkor*. This disease, which threatens the lives of many children in Africa, causes misshapen heads, barrel chests, bloated stomachs, spindly legs and arms, decreased mental abilities, and poor vision.

FATS

Fats, like carbohydrates, are fuel foods, supplying the cells with energy. Certain fats are essential to the structure and function of body cells, to the building of cell membranes, and to the synthesis of certain hormones. Fats also aid in the transport of fat-soluble vitamins. Foods rich in fats include butter, bacon, egg yolk, cream, and certain cheeses.

The energy potential of food is measured in Calories. The Calorie is the unit of heat necessary to raise 1 liter of water 1 degree Celsius. (The large Calorie is always written with a capital C). Fats are concentrated sources of energy. One gram of fat provides nine Calories, while one gram of protein or carbohydrate provides only four Calories. Therefore, foods rich in fats add to the caloric content of the human diet. Some body fat is necessary to cushion body organs and to prevent heat loss through the body's surface. Excessive fat intake causes overweight.

FIBER

The fiber in the human diet comes only from plant sources. Fiber is not a nutrient, but it is important in the diet to stimulate the normal action of the intestines in the elimination of wastes. Fiber absorbs many times its weight in water and aids in the formation of softer stools. Fiber provides bulk which promotes regularity and more frequent elimination.

Currently, it is suggested that dietary fiber may contribute protection against many noninfectious diseases of the large intestine, such as cancer of the colon, hemorrhoids, appendicitis, colitis and diverticulosis. Incidences of these diseases seem to be much lower in countries where the diets are high in fiber. It is also believed that increased dietary fiber reduces blood cholesterol levels and helps to prevent the formation of fatty deposits on the inner walls of the arteries.

Raw fruits and vegetables, whole cereals and bread, and fruits with seeds (strawberries, figs, raspberries) are excellent sources of fiber.

Micronutrients

VITAMINS

The study of micronutrients and their effects on the body was begun in 1906 by Dr. Frederick Gowland Hopkins, a physiology professor at Cambridge University, England. Dr. Hopkins did not isolate these microfactors in food, but he was able to demonstrate serious effects of deficiency diets on white laboratory rats. These illnesses had been noted many times over in human populations. In 1921, Casimir Funk, a Polish scientist, attached the name "vitamine" to these micro food substances which, when missing, caused human illness and body disorders.

Vitamins are organic compounds. They are classified as *water soluble* or *fat soluble*. In general, the water soluble vitamins are coenzymes necessary to the proper sequence of biochemical events that occur during cellular respiration. It is interesting to note that the primates (*Homo sapiens* included) and guinea pigs are the only vertebrate animals that cannot synthesize their own vitamin C from carbohydrates. Therefore, the daily requirements of ascorbic acid must be met through food intake. The functions of the fat soluble vitamins are not clearly understood.

Table 12.1 reviews the major vitamins, their functions and food sources, and the symptoms that commonly result from a deficiency.

TABLE 12.1. Vitamins and Their Uses

Vitamin	Necessary For	Deficiency Symptoms	Food Sources
A-Retinol (fat soluble)	Healthy visual pigments in eye; healthy skin membranes	Dryness of membranes; poor growth; night blindness; inflamed eyelids	Fish liver oil, butter, cream, milk, margarine, brightly colored fruits and vegetables, leafy vegetables
C-Ascorbic Acid[1] (water soluble)	Intercellular cement for teeth and bones; healthy capillary walls; resistance to infection	Sore gums, tendency to bruise easily, painful joints, loss of weight—all these symptoms are associated with scurvy.	Citrus fruits, tomatoes, cabbage, green leafy vegetables, green peppers
D-Calciferol[2] (fat soluble)	Regulates calcium and phosphorus metabolism and growth, building strong bones and teeth	Soft bones, poor tooth development and dental decay—rickets	Fish liver oil, irradiated feed, egg yolk, salmon

TABLE 12.1.   Vitamins and Their Uses (*Continued*)

Vitamin	Necessary For	Deficiency Symptoms	Food Sources
E-Tocopherol	Prevention of oxidation by red blood cells; muscle tone	Hemolysis of red blood cells	Wheat germ, green leafy vegetables
K-Menadione	Synthesis of prothrombin, clotting of blood	Prolonged bleeding from wound	Green vegetables, tomatoes
All the vitamins below belong to vitamin·B complex:			
Thiamin and Niacin[3]	Growth; healthy digestion; normal nerve function; good appetite; carbohydrate metabolism	Poor digestion; depression; nerve disorders; loss of appetite	Yeast, wheat germ, liver, enriched foods, bread, green vegetables
Riboflavin (water soluble)	Growth; health of skin and mouth; functioning of eyes; carbohydrate metabolism	Retarded growth; sores at corner of mouth; disturbances of vision; inflammation of the tongue	Same as for thiamin and niacin; meat

1. Vitamin C is not stored by body, oxidizes rapidly and is readily destroyed by exposure to air.
2. Humans make own vitamin D when skin is exposed to sunlight.
3. Prolonged deficiency of thiamin results in beri-beri, a nerve disease that may result in paralysis. Prolonged deficiency of niacin results in pellagra, a disease characterized by a rash, graying and falling out of hair, depression, loss of weight and digestive disturbance.

MINERALS

Minerals are inorganic compounds. Some such as calcium and sodium are needed in relatively large amounts. Calcium, together with phosphorus, is used in building bones and teeth. Calcium is also a regulator of muscle activity. Nerve cells could not carry impulses nor could muscles contract without the assistance of sodium and potassium. Sodium also functions in the regulation of body temperature, since large amounts of its salts are excreted by the sweat glands. Other minerals are needed by the body in only small amounts. These are known as the *trace* minerals. In general, the functions of trace minerals are regulatory in that they enable the enzymes of metabolism to work.

Table 12.2 reviews the minerals, their function and major food sources, and the symptoms that commonly result from their lack.

TABLE 12.2. Minerals and Their Uses

Mineral	Necessary For	Deficiency Symptoms	Food Sources
Magnesium	Healthy bones and teeth; involved in protein metabolism	Weakening of bones and teeth; faulty metabolism	Green leafy vegetables
Sodium	Functioning of sodium-potassium pump; regulates water balance in cells; regulates nerve impulse; maintains acid-base balance of tissue fluids and blood	Leads to cardiovascular diseases and disorders of the nervous system	Salt
Iron	Synthesis of hemoglobin, myoglobin and the cytochromes	Anemia, difficulties in cellular respiration	Liver, red meats, egg yolk, whole grain cereals
Iodine	Synthesis of thyroxin	Goiter, sluggish metabolism	Marine fish, iodized salts
Fluorine	Aids in resistance to tooth decay	Breakdown of tooth enamel	Water treatment
Calcium	Building of bones and teeth; muscle contraction; nerve impulse transmission; permeability of cell membrane; activation of ATP enzymes	Loss of minerals from bone, anemia, nerve and muscle disorders	Milk and dairy products, eggs, whole grain cereals, green leafy vegetables
Potassium	Functioning of sodium-potassium pump; regulation of nerve impulse; muscle function; glycogen formation; protein synthesis	Nerve and muscle disorders; irregular heart beat	Beans and peas, fruits, vegetables
Phosphorus	Building of bones and teeth; phosphorylation of glucose; building of ATP molecules; functions in cellular respiration; present in nucleic acids	Malfunctions of basic cell processes	Milk and dairy products

Chronology of Famous Names in Biology

1890   **Theodore Palm** (English)—wrote a treatise on the absence of bone deformity (rickets) in poor Japanese children.

1897   **Christian Eijkman** (Indian)—discovered that chickens and people kept on a diet of polished rice developed the disease beriberi.

1898   **J. Lind** (English)—discovered that a diet containing citrus fruits prevents scurvy.

1906   **Frederick G. Hopkins** (English)—applied research methods in inducing and curing deficiency diseases in rats.

1919   **Kurt Huldschinsky** (German)—discovered that ultraviolet radiation and the hormone calciferol can cure rickets.

1920   **E. V. McCollum** and **Lafayette Mendel** (American)—identified the first vitamin and named it A.

1920   **Joseph Goldberger** (American)—discovered that pellagra is caused by a vitamin deficiency.

1921   **Casimir Funk** (Polish)—invented the name "vitamine" to describe a special group of nutrients.

1922   **E. V. McCollum** (American)—discovered vitamin D.

1924   **Harry Steenbock** and **A. F. Hess** (American)—discovered that irradiation of milk increases the vitamin D content.

1950   **D. M. Hadjimarkos** (American)—researched the roles of micronutrients such as selenium, vanadium and barium in the formation of dental caries.

1963   **Marcel El Conrad** (American)—completed original research which elucidated the pathways of iron metabolism in the human body.

1964   **Reginald R. W. Townley** (Tasmanian)—researched carbohydrate malabsorptive disorders in children.

1965   **David Paige** (American)—published a major study on lactose intolerant populations showing lactose enzyme deficiency as the genetic cause for milk intolerance.

1968   **Jana Parizkova** (Czech)—completed research on problems of obesity which included studies of fat deposition and metabolism.

Words for Study

balanced diet

deficiency disease

fiber

glycogen

kwashiorkor

macronutrient

malnutrition

micronutrient

nutrient

nutrition

roughage

vitamin

Questions for Review

PART A. **Completion.** Write in the word that correctly completes each statement.

1. The molecules in food that are used by cells for energy, growth and repair of tissue are ..1..

2. The three major types of macronutrients are carbohydrates, proteins and ..2..

3. Fibrous structures such as hair and nails, enzymes, and hemoglobin all contain the macronutrient ..3..

4. The energy potential of food is measured in ..4..

5. Fiber stimulates the action of the ..5..

6. Some fatty acids are used to synthesize chemical messengers known as ..6..

7. Vitamins and minerals belong to a class of substances known as ..7..

8. Muscle activity is regulated by the mineral ..8..

9. In general the water soluble vitamins are necessary for the events of cellular ..9..

10. Minerals necessary only in small amounts are known as ..10.. minerals.

PART B. **Multiple Choice.** Circle the letter of the item that correctly completes each statement.

1. Which of the following are micronutrients?
   (a) calcium and ascorbic acid
   (b) protein and sucrose
   (c) maltose and calcium
   (d) glycerol and copper

2. Another name for animal starch is
   (a) glycerol
   (b) glycogen
   (c) glucose
   (d) glucagon

3. Which of the following are protein molecules?
   - (a) levulose and vitamin E
   - (b) copper and mannose
   - (c) nitrogen and starch
   - (d) insulin and antibodies

4. The role of fats in the body is
   - (a) functional but not structural
   - (b) structural but not functional
   - (c) functional and structural
   - (d) only structural

5. Fiber in the diet
   - (a) causes diverticulosis
   - (b) prevents infectious diseases
   - (c) increases blood cholesterol
   - (d) absorbs large amounts of water

6. The name Casimir Funk is most correctly associated with
   - (a) the discovery of rickets
   - (b) inventing the name "vitamine"
   - (c) finding the cause of beriberi
   - (d) inventing the term "roughage"

7. The mineral sodium plays a role in all of the following except
   - (a) regulation of body temperature
   - (b) regulation of water balance
   - (c) transmission of nerve impulses
   - (d) synthesis of hemoglobin

8. Prolonged bleeding from a wound can result from a deficiency of
   - (a) vitamin A
   - (b) vitamin D
   - (c) vitamin K
   - (d) vitamin E

9. Vitamins belonging to the B complex group include
   - (a) tocopherol and niacin
   - (b) thiamine and niacin
   - (c) ascorbic acid and menadione
   - (d) calciferol and thiamine

10. Goiter is caused by a deficiency of
    - (a) fluorine
    - (b) zinc
    - (c) iodine
    - (d) magnesium

PART C. Modified True-False. If a statement is true, write "true" for your answer. If a statement is incorrect, change the *underlined* word to one that will make the statement true.

1. Kwashiorkor is a carbohydrate deficiency disease.

2. Nightblindness is caused by lack of vitamin D.

3. The mineral essential to the formation of hemoglobin is zinc.

4. Guinea pigs and *Homo sapiens* can synthesize Vitamin C.

5. Fats are the most abundant of the organic compounds in body cells.

6. The major fuel foods are carbohydrates and <u>proteins</u>.

7. Raw fruits and vegetables are excellent sources of <u>fiber</u>.

8. Rickets is associated with a lack of vitamin <u>C</u>.

9. The mineral <u>phosphorus</u> is necessary for the synthesis of the hormone thyroxin.

10. <u>Fluorine</u> aids in resistance to tooth decay.

Answers to Questions for Review

PART A

1. nutrients
2. fats
3. protein
4. Calories
5. intestine
6. hormones
7. micronutrients
8. calcium
9. respiration
10. trace

PART B

1. a
2. b
3. d
4. c
5. d
6. b
7. d
8. c
9. b
10. c

PART C

1. protein
2. A
3. iron
4. cannot
5. Proteins
6. fats
7. true
8. D
9. iodine
10. true

CHAPTER 13

THE DISEASES OF *HOMO SAPIENS*

A disease is a disorder that prevents the body organs from working as they should. In general, diseases can be classified as being *infectious* or *noninfectious*. Infectious diseases are caused by organisms that invade the body and do harm to the cells, tissues and organs. As a rule, there is disease specificity whereby a specific disease-producing organism causes a particular disease. Disease-producing organisms are said to be *pathogens* and are described as being *pathogenic*. Usually, pathogens—or germs, as they are often called—are microorganisms, organisms microscopic in size. Microorganisms pathogenic include certain bacteria, protozoans, spirochetes, richettsias, mycoplasmas, and fungi. Parasitic worms and viruses also often produce disease in humans. Most infectious diseases are *contagious*—capable of being passed from one person to another by means of body contact or by droplet infection.

Noninfectious diseases are caused by factors other than pathogenic organisms. Among the factors that are responsible for noninfectious diseases are genetic causes, malnutrition, exposure to radiation, emotional disturbances, organ failure, poisoning, endocrine malfunctioning and immunological disorders. Whatever the cause, a disease works counter to the well being of a human.

Infectious Diseases

METHODS OF SPREADING

Contagious or *communicable* diseases are spread from one person to another in a number of ways. One such way is *direct contact* whereby the infected person touches a well person. This may be accomplished by handshake, kissing, or through sexual intercourse (venereal diseases). Microorganisms are also spread *indirectly* when a noninfected person handles objects that have been in contact with the infected person. Eating from the same plate, drinking from the same glass, handling bed clothes or towels or any number of items are means by which germs are passed.

Droplet infection is a common method of passing germs along. Disease germs are present in droplets of water that escape from the nose and mouth when sneezing, coughing and talking. If these infected droplets are inhaled or taken in by mouth, the germs then enter the body of another person. Animals may be vectors, or carriers, of diseases. Biting insects or mammals may carry disease germs in their salivary glands and pass them on to a human who is bitten. Animal hair carries insects which may be disease carriers. Contaminated food and water spread diseases throughout human populations. Finally, human carriers of disease organisms (who remain unaffected by the germs that they carry) can spread pathogenic organisms to noncarriers.

COMMON INFECTIOUS DISEASES AND THEIR CAUSES

Bacteria are responsible for many human diseases, as outlined in Table 13.1.

Certain rickettsia can also produce disease in humans when they enter the body through the bites of their hosts—mites, ticks, lice, and fleas. Rickettsia diseases are often serious, characterized by high fever and rash, and often lead to death (Table 13.2). Some protozoa and some worms can also produce disease in humans. Most parasitic worms are ingested in contaminated food, usually encysted in the muscles of cows, pigs, sheep, and snails (Table 13.3).

A few species of yeasts and molds are pathogenic for humans. They attack the skin, mucous membranes and the lungs. Ringworm and athlete's foot are two such diseases. Although irritating and inconvenient, these diseases are not generally serious and can usually be readily treated.

Viruses—nonliving particles that come "alive" when they invade a cell—are major disease-producers in humans. Some viruses cause disease that affects the entire body; others cause disorders that affect a particular organ. Table 13.4 lists some common viral diseases.

TABLE 13.1. Diseases Caused by Bacteria

Diseases Caused by Bacilli (Rods)	Diseases Caused by Cocci (Spheres)	Diseases Caused by Spirellae (Spirals)
Tuberculosis	Pneumonia (some forms)	Syphilis
Diphtheria	Gonorrhea	Asiatic Cholera
Tetanus	Scarlet Fever	Yaws
Typhoid Fever	Rheumatic Fever	
Bubonic Plague	Streptococcus sore throat	
Whooping Cough	Meningitis (some forms)	
Tuleremia	Childbed fever	
Leprosy		

TABLE 13.2.   Rickettsia Diseases

Human Disease	Disease Vector
Typhus	Lice
Trench Fever	Lice
Rocky Mountain Spotted Fever	Ticks
Rickettsial Pox	Mites

TABLE 13.3.   Worm-Caused Diseases

Diseases Caused by Flatworms	Diseases Caused by Roundworms
Tapeworm Infection	Hookworm
Sheep Liver Fluke Infection	Trichinosis
Chinese Liver Fluke Infection	Pinworm Infection
	Ascaris Infection
	Filariasis

TABLE 13.4.   Common Virus Diseases

Virus Disease	Organ Affected
Poliomyelitis	Nervous system and muscles
Rabies	Nervous system
Encephalitis	Nervous system
Viral Pneumonia	Lungs
Common Cold	Respiratory system
Influenza	Lungs and respiratory system
Fever blisters	Skin around the lips
Genital Herpes	Genital organs
Mumps	Salivary glands
Viral Hepatitis	Liver
Trachoma	Eyes
Measles	
Smallpox	
Chicken pox	
Yellow fever	spread throughout the body
Dengue fever	
Psittacosis (parrot fever)	

HOW PATHOGENS DAMAGE THE BODY

Once pathogenic organisms enter the body there is interaction between the body and the germs that results in disease. The type of disease is determined by the type of pathogen that invades the body. The severity of the disease depends on the ability of the body to ward off infection (*resistance*) and the strength or *virulence* of the infecting germ.

Pathogens may affect body tissues and functions in a number of ways. Some pathogens produce enzymes that dissolve the materials that hold cells together, creating pathways for germs to enter tissues. Other germs produce substances that kill certain body cells. Pathogens frequently damage only certain cells and tissues. The rickettsia of Rocky Mountain spotted fever damages the liver as does the Chinese liver fluke. The polio virus destroys nerve cells. The spirochete of syphilis often destroys brain tissue. The typhoid bacillus attacks the lymph tissue of the intestinal wall.

Some disease-producing microorganisms secrete *toxins* that interfere with the metabolic activities of cells. Certain germs block vital passageways of the body and thus prevent normal functioning; for example, the organisms that cause diphtheria seal the throat with membranes and prevent breathing. Worm parasites frequently compete with the host for nutrients and produce malnutrition in the host.

BODY DEFENSES AGAINST DISEASE

When the body is attacked by pathogenic organisms it puts up a series of defenses, designed to destroy the enemy and maintain health.

First Line—Skin and Outer Defenses

The first line of defense against invasion of the body by germs is the skin. The clean, unbroken skin is thick enough and tough enough to prevent most germs from penetrating. As a rule, germs that land on the skin do not live long enough to cause trouble because the skin itself has a germicidal quality that inhibits the growth of germs on its surface.

The eyes, nose and mouth are in effect breaks in the skin that could permit the penetration of germs into the body. Most germs entering the eye do not live long enough to cause distress. They are dissolved by *lysozyme*, an enzyme in tears. Nevertheless, some virulent germs survive and produce eye infections such as conjunctivitis (pink eye) or *tracoma*. Tracoma, a virus disease, is especially dangerous because it often causes blindness.

Germs numbering in the thousands enter the mouth daily with food and drink. Few of these survive to reach the intestines. The saliva in the mouth is able to kill many of the invaders. Those that reach the stomach face the killing action of hydrochloric acid and the digesting power of pepsin. However, some germs manage to survive. The germs of Asiatic cholera, typhoid, paratyphoid and other serious intestinal diseases are able to resist these body defenses and cause illness.

Uncountable numbers of disease germs are breathed in through the nose from the surrounding air. However, few of these ever reach the lungs. The nasal passages present a complicated maze of filters guarded by hairs that trap many germs. In addition, the mucus membranes lining the air passages secrete sticky mucus which traps disease germs, rendering them inactive. Sneezing forces germs to the outside, also.

Germs that manage to reach the breathing tubes are for the most part trapped in mucus secretions from the cells that compose these tubes. In addition, the cilia of the cells that line the air tubes sweep the mucus-trapped germs back to the throat where they are swallowed and then destroyed in the stomach by hydrochloric acid and pepsin. Special "dust cells" in the air sacs of the lungs pick up some germs and carry them out. Despite these active defenses, some germs survive and cause respiratory diseases such as colds, influenza and pneumonia.

Second Line—Confinement

Despite the efficient outer defenses, some germs manage to enter the body. When germs break through the skin, a series of body activities occur which are attempts to ward off the spread of these germs through the body. When tissue damage begins, certain chemicals are released that cause the local capillaries to expand. These blood vessels become filled with blood, producing redness and local fever. Plasma seeps out of the capillaries into the tissue spaces, causing local swelling. Pressure on nerve endings in the area produces pain and tenderness. All of this activity is accompanied by the arrival of hordes of phagocytic white blood cells that force their way through the capillary walls into the tissue spaces outside where they try to devour the germs through phagocytosis.

Meanwhile the fibrinogen in the plasma solidifies into a tangled mass of strands and fibers. Germs trying to slip through are trapped. New fibrous cells develop in and around the entire affected area. This tends to wall off the danger point and prevent the germs from spreading. Any boil or abscess is visible proof that this walling-off process localizes an infection. If all of this is successful, the germs are confined to one small area, and wiped out by the phagocytic white blood cells. The damaged tissues are repaired and soon the infected area is as good as new.

Third Line—Macrophage Activity

Certain pathogens can resist the phagocytic activities of white blood cells. These germs may be enclosed in a slimy capsule or they may be surrounded by flagella that whip back and forth. Such protected germs are hard to pin down and engulf. Still other germs produce chemicals that are distasteful or even poisonous to white blood cells. If the germs can keep the white blood cells at bay, they can reproduce at leisure and spread.

When the initially responding white blood cells have difficulty in devouring the germs, they are joined by a larger and tougher phagocytic cell known as a *macrophage*. Macrophages are giants among the white blood cells, and they devour germs and white blood cells that have difficulty handling foreign particles. Macrophages are active metabolically. They contain large numbers of lysosomes that possess very powerful hydrolytic enzymes. These enzymes digest both bacteria and other smaller white blood cells.

Fourth Line—Lymph

Lymph also defends the body from disease. You know that lymph is fluid that diffuses out of the blood into the tissue spaces bathing the cells. The lymph is then drained away by a series of fine tubes known as lymph vessels. The lymph vessels are equipped with special nodes or glands that filter the lymph as it passes through them in its travels around the body. Each time lymph enters one of the lymph nodes, bacteria are picked out and gobbled up by special cells. By the time the lymph leaves the gland, it contains fewer bacteria than it did when it entered. As lymph moves through each of these gland filters, the same procedure is repeated. By the time lymph leaves the lymph vessels and is returned to the blood, it is almost free of bacteria.

Fifth Line—Antibodies

The body also produces substances known as *antibodies* to fight disease-producing agents. Antibody production is a relatively slow process. First of all, the body cells must recognize the invading agent as "foreign"— an antigen—and then produce an antibody that is exactly right to immobilize the protein invader. Finally, the blood cells must go into full scale production of this specific antibody. An antibody is specific against each type of germ. For example: diphtheria antibodies will not be effective against scarlet fever antigen.

Immune globulins (also known as *gamma globulins*) are proteins in the blood plasma and body fluids. Specifically, these immune globulins are antibodies that are produced in response to infection or to injection of killed bacteria, foreign proteins or other substances. Currently, five classes of immune globulins have been identified and are designated as IgM, IgG, IgA, IgD, and IgE. Each of these immune proteins is present in serum and body fluids in a characteristic amount. IgG and IgM have the ability to bind and clump invading bacteria; the others work differently.

PROTECTION AGAINST DISEASE

Immunization

Immunity is the ability to resist the attack of a particular disease-producing organism. Immunity to one kind of disease germ does not automatically make a person immune to other types of disease germs. *Active immunity* is brought about by antibody production by a person's own body cells. Active immunity can be stimulated in either of two ways; by getting the disease and recovering from it or by being immunized against the disease. Immunization that produces active immunity involves the injection of weakened disease agents that stimulate antibody production but produce only mild symptoms or none at all. Active immunity is long-lasting because the body cells continue to produce the antibodies.

An injection of gamma globulins can give a person temporary immunity against certain specific diseases. This means that a person has borrowed antibodies in the blood and not those made by his (her) own cells. This kind of immunity is called *passive immunity*. It lasts only as long as the antibodies last; when they are used up, the immunity ceases.

Smallpox is a disease that has been almost entirely eradicated from even remote corners of the earth. The fight against this disease began in the 18th century when Edward Jenner vaccinated people with cowpox. Vaccination with cowpox stimulates the body to build its own antibodies and give lasting immunity to the dread disease of smallpox. Cowpox is a mild disease in humans and stimulates the cells to produce antibodies which happen to be effective against smallpox.

Successful methods of immunization have been developed against diphtheria, polio, whooping cough, lockjaw, plague, typhoid, yellow fever, cholera, measles, mumps, rubella, and typhus. In every community in the United States children are automatically immunized against diphtheria, tetanus, whooping cough, polio and smallpox. Many of the states have laws that prevent children from attending public school without the prescribed immunizations.

Safe Drinking Water

At one time diseases such as typhoid fever and cholera devastated the population of New York. Today not one death can be attributed to either of these diseases. The reason for the marked decrease in the incidences of these diseases is attributed to the development of a safe water supply.

Sanitary water engineers are employed to protect the community against the dangers of contaminated water. They use many water purification techniques. *Settling* is a process in which water is held in large tanks until suspended solids settle out. *Filtering* is accomplished by allowing water to trickle through sand beds several feet deep. This removes 90–95 percent of all bacteria as well as fine particles of solid matter. *Aeration* is the process in which water is sprayed up into the air. This technique kills some bacteria and allows more air to dissolve into the water. *Chlorination* is the chemical purification of water; chlorine gas or hypochlorite is added to the water in small quantities to kill any remaining pathogens.

Sanitary engineers and bacteriologists watch the community water supply very closely. They test the water supply constantly for *E. coli.* The presence of this bacterium is known as the *index of fecal contamination.* Since *E. coli* are normally present in the human intestines, their presence in drinking water indicates that the water supply must be contaminated with sewage. When *E. coli* is not present, the water is free of human wastes and is probably free of organisms that cause typhoid, cholera, and other diseases of the human intestines.

Safe Milk and Food

At one time diseases such as tuberculosis, Q fever, septic sore throat, brucellosis and stomach upsets were transmitted to people from contaminated milk. To prevent the spread of disease through milk most local laws require that milk be *pasteurized* before sales. Commercial dairies chill the milk as quickly as possible after collection. This temporarily inhibits the growth of bacteria which may thrive at the body temperature of the cow. Then the milk is heated for a certain time and at a certain temperature that will kill the toughest disease germs present. This process is called *pasteurization.* After pasteurization, milk still contains thousands of living bacteria, but the disease-producing germs have been destroyed.

Methods of food preservation are used to prevent food from being spoiled and contaminated by bacteria. Canning sterilizes food and seals it up so that no bacteria can get in. Other methods are used that prevent the growth of bacteria in food products and prevent rapid spoilage. Some of these methods were devised by the ancients; some are relatively new. They include drying, salting, sugaring, pickling, fermenting, smoking, refrigeration, fast feeezing, sterilization by heat or radiation, and the addition of chemical preservatives.

Noninfectious Diseases _____

The noninfectious diseases are not communicable because they are not caused by infectious organisms. These diseases have various causes other than germs. Table 13.5 provides a listing of some of the noncommunicable diseases.

Not all of the diseases listed in the table will be discussed in this chapter. Diseases resulting from endocrine insufficiency or oversecretion were summarized in Chapter 11, and some disorders that result from nutritional deficiency were summarized in Chapter 12.

CARDIOVASCULAR DISEASES

Problems of the heart and blood vessels are known as *cardiovascular diseases*. Because diseases of the heart and the circulatory system are leading causes of death in the United States, a great deal of research effort has been put forth for many years to determine the causes of heart defects and to find ways to prevent and/or cure these disorders. Heart defects trouble people of all ages.

Many infants are born with *congenital* heart defects. Sometimes the heart defect is a hole in the heart wall which divides the left and right ventricles. In some cases, the large arteries leaving the heart are in the wrong place. It also happens that a blood vessel that should have closed at birth failed to do so. Today many of these defects in childrens' hearts are corrected through procedures of *open heart surgery*. During the operation, the blood is sent through a machine that serves as a mechanical heart and lungs. After the real heart is repaired, the blood is returned to its normal pathways through the body.

In young and old alike, defective heart valves may cause trouble. Valves that are beyond self repair are replaced by plastic ones.

Many things may go wrong with the adult heart. Sometimes the Purkinje fibers in the heart lose their ability to contract rhythmically and thus the pacemaking ability of the heart is impaired. This defect is helped by implanting an artificial *pacemaker* in the patient's chest which makes possible appropriate heart stimulation.

Coronary artery disease prevents the proper blood supply from reaching the heart. The cells in heart muscle become damaged when not enough oxygen and nutrients reach them. Damaged heart cells cause heart attacks. If the damage to the heart is not too severe, a person will recover. Severe heart attacks cause death. High blood pressure (*hypertension*) is another major cause of heart attacks and strokes.

In recent years, some highly skilled heart surgeons have attempted exotic methods for dealing wih patients who have badly diseased hearts. One of these methods has been heart transplant in which a diseased heart is exchanged for a healthy one. This technique has met with limited success in which the lives of transplant patients have been prolonged for a few months to a year. Recently, a patient was kept alive for about three months by means of an artificial heart. Neither of these methods has proven practical for large numbers of cases.

TABLE 13.5. Some Noninfectious Diseases

Disease Type	Malfunction
Cardiovascular disease	Heart and blood vessels
Allergy	Hypersensitivity to certain substances
Genetic	Inborn defects
Emotional including drug-related psychoses	Psychological problems and mental stress
Occupational	Physical problems caused by work conditions
Poisoning	Illness caused by tissue toxins
Nutritional	Nutrient deficiencies in diet
Hormonal	Imbalance in endocrine secretions
Cancer	Wild reproduction of cells

ALLERGY

Some people are very sensitive to substances that are quite harmless to most other people. The cells of these sensitive people produce antibodies to ward off whatever substance affects them. The antibodies become attached to the tissue cells, rendering the person *sensitized.*

Whenever that particular substance enters the body again, it reacts with the attached antibodies and damages the cells. These damaged cells prompt certain symptoms such as itching, sneezing, tearing eyes, red welts, large hives, fever, and a general feeling of not being well. The injured cells may release chemicals called *histamines* which are responsible for the symptoms. *Anti-histamines* are substances that may neutralize the histamines and relieve the symptoms. People may be sensitive to things that they inhale such as pollen, dust or powders. Other people may be sensitive to things that they eat such as wheat, eggs, milk, fish, nuts, chocolate, strawberries or bananas. There are those people who get violent reactions from drugs such as penicillin, aspirin, or streptomycin. A person may become allergic to a given substance at any time in life.

OCCUPATIONAL DISEASES

The jobs of many people put them in contact with substances that present hazards to health. Fumes, dust, gases, vapors, fibers and chemicals may have devastating effects on the health of a large number of people. Not too long ago it was discovered that asbestos causes lung cancer in those who have regular exposure to its fibers. Similarly, people who inhale the fibers of cotton or sugar cane may develop debilitating lung conditions and breathing problems. *Silicosis* is a type of lung disease occurring in miners and sandblasters who are in constant contact with rock dust.

Some occupations put workers in contact with poisons. Painters used to be subjected to lead poisoning from lead-base paints. (Today, most household paints are made without lead.) Lead is toxic to the body, settles in the brain and damages brain tissue. It is not unusual for farmers and exterminators to become poisoned by pesticides which they may absorb into the body by contact or by breathing. Recently, the pesticide Chlordane was shown to have adverse effects on the health of persons whose homes were contaminated by misuse of the product.

People in the medical professions are subject to occupational hazards. The ionizing radiation produced by X-ray machines can cause a wide variety of unfavorable conditions.

CANCER

Cancer is not a single disease. It is a whole constellation of different diseases that share a common characteristic. In every form of cancer there is an abnormal, uncontrolled growth of body cells. The cells that grow so wildly are not foreign cells that have invaded the body, but regular body cells that somehow have gone wrong. They grow in a disorganized fashion and compete with normal cells for space and nutrition.

These wildly growing cells form a malignant mass called a *neoplasm* or a *cancer*. The malignant mass sends out fingers of cancerous cells that burrow into the normal tissues around it. As the invasion progresses, the normal tissue is gradually destroyed.

In the beginning cancer is localized in a particular part of the body. As time passes, fragments of malignant tissue separate away from the original malignant mass. They are carried off in the blood and lymph, reaching many new sites in the body. Wherever they remain, a new colony is established and grows like the original mass. This process in which cancer spreads is called *metastasis*.

All cancerous cells do not grow at the same rate. Some types of cancer grow quickly; others grow slowly. All cancer ultimately destroys normal tissue and kills the host. Every living cell has the potential to turn cancerous. Every variety of vertebrate animal is subject to the disease. So are humans of every race, nation or habitat. Cancer cannot be blamed on modern living. Cancers have been found in Egyptian mummies and in fossils of ancient dinosaurs.

Many causes for cancer have been suggested. Some of these suggestions are more than guesses or hunches having been based on statistical studies of the incidences of disease. A case in point is cigarette smoking. Rather extensive research work has shown a very definite link between cancer (heart disease, too) and lung cancer. Smokers also show a high incidence of cancer of the lips, mouth and throat.

Research has also indicated that some forms of cancer seem to run in families. For example: the daughters and sisters of women with breast cancer, and the children of patients with rectal cancer should be informed of their increased risk. There seems to be familial patterns of occurrence in these diseases.

There is strong evidence that many types of chemicals are *carcinogens*, or cancer producers. Contact over a period of time with certain defoliants, insecticides and "buried" chemical wastes seems to have caused cancers of multiple varieties in large segments of the population. It is known that radium workers tend to get bone cancer. Workers in factories that produce aniline dyes often get cancer of the bladder.

Homo sapiens is constantly bombarded by various kinds of radiation: ultraviolet rays, X rays, gamma rays, cosmic rays, radioactive fallout and the like. Any form of radiation, natural or human-made, can produce leukemia, bone cancer, skin cancer or other types of cancer, such as lung, breast or thyroid cancers. The effect of the atomic bomb explosions on Hiroshima and Nagasaki in Japan leave no doubt that a single exposure to a high dose of radiation can produce leukemia.

There is some indication that certain forms of cancer may be communicable, transmitted from one person to another by way of a virus. For example: cancer of the cervix may be initiated by infection with herpes virus. These virus particles are found commonly in cancerous cervix cells. It is also known that the second wife of a man whose first wife died from cancer of the cervix usually develops cancer of the cervix also. The spread of Kaposi's sarcoma (a rare form of cancer associated with AIDS, or autoimmune deficiency syndrome) through the homosexual population may indicate viral transmission. The electron microscope has revealed that some leukemia cells contain virus particles.

The best protection against any type of disease is to avoid contact with causative agents or chemicals whenever possible.

Chronology of Famous Names in Biology

460 BC	**Hippocrates** (Greek)—"father of medicine" because he was the first of the ancients to attempt scientific explanations of disease.
1527	**Jacques de Bothencourt** (French)—gave the name "venereal" to those diseases transmitted by sexual intercourse.
1530	**Girolamo Fracastoro** (Italian)—gave the name "syphilis" to the heretofore unnamed venereal disease.
1796	**Edward Jenner** (English)—discovered that vaccination with cowpox renders immunity against smallpox.
1854	**Louis Pasteur** (French)—proved that microorganisms caused fermentation. In 1885 he developed the treatment for rabies.
1860	**Philippe Ricord** (American)—determined that gonorrhea and syphilis are two separate diseases.
1879	**Albert Neisser** (German)—isolated the organism that causes gonorrhea.
1872	**Mortiz Kaposi** (Rumanian)—first described the symptoms of the type of cancer that now bears his name.
1880	**Charles Laveran** (French)—described the protozoan parasite of malaria.
1882	**Elie Metchnikoff** (Russian)—discovered the phagocytic activities of white blood cells.

1883   **Robert Koch** (German)—discovered the bacterium that causes tuberculosis.

1892   **Walter Reed** (American)—discovered that yellow fever was transmitted by Aedes mosquito.

1896   **Joseph Lister** (English)—developed aseptic techniques to prevent infection during and after surgery.

1906   **Howard T. Ricketts** (American)—discovered that ticks infected with microorganisms (Rickettsia) cause Rocky Mountain spotted fever.

1908   **Paul Ehrlich** (German)—developed the first chemical substance—salvarsan—to inhibit the growth of the syphilis spirochete.

1911   **Emil von Behring** (German)—discovered that immunity to diphtheria could be given by injecting a person with antitoxin.

1929   **Alexander Fleming** (English)—discovered the growth inhibitions effect of penicillin on staphylococcus.

1932   **Gerhard Domagk** (German)—discovered that *prontosil* is effective against Streptococcus infections.

1933   **George Dick** (American)—devised a test to determine susceptibility to scarlet fever and discovered the germ that causes that disease.

1933   **Bela Schick** (American)—developed the Schick test for susceptibility to diphtheria.

1935   **Wendell Stanley** (American)—isolated the virus that causes tobacco mosaic disease.

1943   **Howard Florey** (American)—developed an efficient method of mass producing penicillin from the mold *Penicillium notatum*.

1944   **Peter Medawar** (English)—did the basic research which brought to light the problems of tissue and organ transplant techniques and immunology. 1967—discovered acquired immune tolerance factors that prevent tissue transplants.

1948   **Selman Waksman** (American)—first to isolate streptomycin from *Streptomyces griseus*

1954   **Jonas Salk** (American)—developed a vaccine effective against poliomyelitis.

1963   **Albert Sabin** (American)—developed oral polio vaccine.

1965   **K. Ishizaka** (American)—discovered gamma globulin E (IgE) which is present in normal serum in small amounts.

Words for Study

active immunity	lymphocytes
allergy	macrophage
antibody	metastasis
antigen	neoplasm
anti-histamine	passive immunity
aeration	pasteurization
cardiovascular diseases	resistance
contagious	sarcoma
droplet infection	settling
genetic diseases	tracoma
histamine	tubercle
hypertension	vaccine
immune globulins	virulence
immunization	venereal disease
index of fecal contamination	

Questions for Review

PART A.   Completion. Write in the word that correctly completes each statement.

1. Diseases that are caused by viruses, bacteria or other pathogens are known collectively as ..1.. diseases.
2. A disease that is spread from one person to another is said to be ..2..
3. Rocky Mountain spotted fever is caused by the organism called a ..3..
4. Polio is caused by ..4.. infection.
5. Most parasitic worms enter the body by way of contaminated ..5..
6. The disease-producing ability of a pathogen is summed up by the term ..6..
7. Tracoma is a disease of the eye caused by a ..7..
8. Phagocytosis is best associated with ..8.. blood cells.
9. Fluid that bathes the body spaces is called ..9..
10. Immune blood proteins are the ..10.. globulins.
11. The ability to resist disease is known as ..11..
12. Cholera is spread through unclean ..12..
13. Milk is ..13.. to prevent the spread of tuberculosis and Q fever.
14. The work of the Purkinje fibers can be taken over by an artificial ..14..
15. During allergic attacks, cells release chemicals known as ..15..

PART B. Multiple Choice. Circle the letter of the item that correctly completes each statement.

1. The pathogens that are not microscopic in size are
   (a) virus particles   (c) worms
   (b) bacteria   (d) mycoplasmas

2. Transmission of germs by direct contact may be accomplished by
   (a) droplet infection   (c) handling clothing
   (b) handshake   (d) sneezing

3. An example of a disease caused by a bacillus is
   (a) meningitis   (c) syphilis
   (b) yaws   (d) tetanus

4. Viruses reproduce
   (a) in living cells   (c) in quiet waters
   (b) on dead organic matter   (d) in blood plasma

5. Ringworm is a disease of the skin that is caused by infection with
   (a) protozoa   (c) hookworm
   (b) fungus   (d) bacteria

6. When germs break through the skin, certain chemicals are released from body cells that
   (a) kill the germs immediately   (c) cause the capillaries to expand
   (b) seal up the wound   (d) prevent pain and tenderness

7. Microorganisms that are enclosed in capsules are usually
   (a) phagocytic   (c) harmless
   (b) anaerobic   (d) pathogenic

8. By the time lymph leaves the lymph vessels and is returned to the blood, the lymph is
   (a) absolutely sterile   (c) almost free of bacteria
   (b) able to destroy bacteria (d) crowded with bacteria

9. An injection of gamma globulins can give a person the type of immunity best described as
   (a) partial   (c) lasting
   (b) temoporary   (d) active

10. The purpose of the aeration of water is to
    (a) remove debris   (c) prevent tooth decay
    (b) improve the color   (d) kill anaerobes

11. Allergy is caused by sensitizing
    (a) lymphocytes   (c) leucocytes
    (b) antibodies   (d) antitoxins

12. A malignant mass of cells is known as a (an)
    (a) metastasis   (c) ectoplasm
    (b) protoplasm   (d) neoplasm

13. Carcinogens are
    (a) virulent pathogens   (c) contact poisons
    (b) immune proteins   (d) cancer producers

14. A disease vector
    (a) carries the disease
    (b) causes the disease
    (c) cures the disease
    (d) controls the disease

15. The phagocytic activities of white blood cells were discovered by
    (a) Jenner
    (b) Metchnikoff
    (c) Salk
    (d) Sabin

PART C.   Modified True-False. If a statement is true, write "true" for your answer. If a statement is incorrect, change the underlined word to one that will make the statement true.

1. Malnutrition is an example of a <u>contagious</u> disease.

2. Human carriers <u>are</u> affected by the germs they carry.

3. Leprosy is a disease that is caused by a <u>coccus</u>.

4. Rickettsia are microorganisms that live in the body of <u>protozoa</u>.

5. Smallpox is a <u>bacterial</u> disease that affects the whole body.

6. Hepatitis is an infection of the <u>liver</u>.

7. The ability of the body to ward off infection by disease organisms is known as <u>virulence</u>.

8. The virus of polio destroys <u>lung</u> tissue.

9. Lysozyme is <u>a hormone</u> in tears.

10. Phagocytic white cells <u>egest</u> bacteria.

11. A boil is the result of a <u>spreading</u> infection.

12. Macrophages are very <u>small</u> white blood cells.

13. The name, Edward Jenner, is best associated with the disease <u>polio</u>.

14. Cowpox provides <u>passive</u> immunity.

15. Methods of food preservation include salting, freezing and <u>drying</u>.

Answers to Questions for Review

PART A

1. infectious
2. communicable or contagious
3. rickettsia
4. virus (viral)
5. food
6. virulence
7. virus
8. white
9. lymph
10. gamma
11. immunity (resistance)
12. water
13. pasteurized
14. pacemaker
15. histamines

PART B

1. c	6. c	11. c
2. b	7. d	12. d
3. d	8. c	13. d
4. a	9. b	14. a
5. b	10. d	15. b

PART C

1. noninfectious
2. are not
3. bacillus
4. ticks, mites, lice, fleas
5. virus
6. true
7. resistance
8. nerve or muscle
9. enzyme
10. engulf or ingest
11. local (contained)
12. large
13. smallpox
14. active
15. true

CHAPTER 14

HEREDITY AND GENETICS

The science of genetics has made tremendous advances since the 1960's but the foundations lay at the beginning of the 20th century. Before discussing the modern findings of genetics we should review the classical principles upon which the science is founded.

Classical Principles of Heredity

MENDEL'S LAWS OF HEREDITY

Gregor Mendel, an Austrian monk, began the first organized and mathematical study of how traits are inherited. Using the garden pea as the test organism, Mendel identified seven different traits which were easily recognizable in this self-pollinating plant. He called each of these traits *unit characters*. Mendel began his work in 1856 when little was known about chromosomes and their functions in cell division and the concept of the gene had not yet been developed.

Mendel not only identified characteristics that seemed to be inherited, but for each unit character, he identified an opposite trait. For example: if the unit character was height, the opposite traits were short and tall. If the unit character was seed coat color, the opposite traits were yellow and green.

Based on his observations of crosses that he made in pea plants by means of hand pollination, Mendel formulated three major laws or principles of inheritance. It should be noted that he kept very careful records of his experiments and converted his results into mathematical ratios.

The Law of Dominance

If two organisms that exhibit contrasting traits are crossed, the trait that shows up in the first filial generation (F_1) is the dominant trait. For example: when a pure-bred tall pea plant is crossed with a short pea plant, all of the offspring will be tall. The offspring will not be pure tall, however, and are therefore known as *hybrids*. The factor for shortness is hidden. We say today that the *phenotype* of the F_1 plants is tall. A phenotype refers to the traits that we can see. The *genotypes* or genetic makeup of these plants is said to be hybrid or *heterozygous*, meaning mixed.

When organisms with contrasting traits are crossed, the trait that shows up in the F_1 generation is called the *dominant* trait. The trait that is hidden is called the *recessive* trait. Since Mendel did not have the concept of the gene, he called the conditions that make for dominance and recessiveness factors. From the results of crosses made with the garden pea, he concluded that (a) two factors determine a characteristic and that (b) two recessive factors are needed in order for the recessive characteristic to show up and (c) that one dominant factor and one recessive factor produce offspring that have the dominant trait.

The Law of Segregation

When hybrids are crossed, the recessive trait segregates out at a ratio of three individuals with the dominant trait to one individual with the recessive trait. The 3:1 ratio is known as the *phenotypic* ratio, because it refers to the traits that can be seen and not those factors hidden in the germplasm. The hybrid cross is also known as the F_1 cross and the offspring produced by this cross are known as the second filial generation, or F_2. In terms of modern knowledge, the F_2 generation also produces another type of ratio called the *genotypic ratio* which refers to *gene* makeup. The genotypic ratio is 1:2:1, translated into 1 *homozygous* dominant (pure) individual: 2 *heterozygous* (hybrid) individuals: 1 *homozygous* recessive. Only the homozygous recessive shows the recessive trait.

The Law of Independent Assortment

Mendel believed that each trait is inherited independently of others and remains unaltered throughout all generations. We now know that Mendel's "factors" are genes that are linked together on chromosomes and that if genes are on the same chromosome, they are inherited together.

THE CONCEPT OF THE GENE

Around 1911, Thomas Hunt Morgan introduced the tiny fruit fly *Drosophila melanogaster* as the new experimental organism for work in the field of heredity. The experimental work of Morgan resulted in the discovery that the chromosome is the means by which hereditary traits are transmitted from one generation to another. Morgan's chromosome theory of inheritance includes the concept that chromosomes are composed of discrete units called *genes*. Genes are the actual carriers of specific traits and move with the chromosomes in mitotic and meiotic cell divisions. Morgan further proposed that genes control the development of traits in each organism. When genes change, or *mutate*, the traits they control change.

SOME GENETIC SHORTHAND

A combination of the theories of Mendel and Morgan led to the development of a system of genetic "shorthand" which is used to represent the genetic makeup of a trait and to show rather simply what happens when organisms with specific traits are crossed. In this shorthand capital letters are used to indicate dominant genes, lowercase letters for recessive genes. A Punnett square is a diagrammatic device used to predict the genotypic and phenotypic ratios that will result when certain gametes fuse. Remember that as a result of meiosis each gamete has only one half the number of chromosomes that are in the somatic cells.

Problem: In fruitflies, long wing (L) is dominant over vestigial wing (l). What is the result of a cross between two flies that are heterozygous (Ll) for wing length?

Solution:

Parents: Male   ×   Female
                  Ll Ll

Gametes: Ⓛ Ⓛ   Ⓛ Ⓛ

Punnett square

	L	l
L	LL	Ll
l	Ll	ll

F_2 LL = 1 homozygous dominant long winged fly
          Ll = 2 heterozygous dominant long winged flies
          ll = 1 homozygous recessive short winged fly

INTERMEDIATE INHERITANCE

Geneticists have discovered that in many cases a trait is not controlled by a single gene, but rather by the cooperative action of two or more genes. There are many instances in which Mendel's "law of dominance" does not hold true. A case in point is what was once called *blending inheritance,* or *incomplete dominance.* It is now known as *codominance.* When red-flowered evening primroses are crossed with white-flowered primroses, the hybrids are pink. Neither red nor white color is dominant and therefore the result is a blend. In sweet peas, the expression of red or white flowers is dependent upon two genes: a (C) gene for color and an (R) gene for enzyme. If C and R are inherited together, the flower color is red. If the dominant C is missing, the flower is white and if the dominant R is missing, the color is white. Therefore white flowers are the result of several different genotypes: *ccrr, ccRR, CCrr, Ccrr.*

CROSSING OVER

Genes are linked on chromosomes and are inherited in a group on a particular chromosome. However, linkage groups are broken by *crossing over,* a phenomenon which may occur during meiosis when homologous chromosomes are intertwined during synapsis. It is at this time that chromosomes may exchange homologous parts and thus assort linkage groups.

MUTATIONS

It was stated before that genes can change and that changes in genes are known as *mutations*. As a rule, mutations are usually recessive and they are usually harmful. Mutations usually occur at random and spontaneously. However, mutations may be induced by radiation or by chemical contamination.

There are several types of mutations. A loss of a piece of a chromosome is known as a *deletion*. The genes on the broken off piece of chromosome are lost. Sometimes a broken piece of chromosome sticks on to another chromosome, thus adding too many genes; this type of mutation is known as *duplication*. Sometimes a piece of chromosome becomes rearranged in the chromosome where it belongs, thus changing the sequence of the genes on that chromosome; this is known as an *inversion*, and it prevents gene for gene matching when chromosomes line up during meiosis. *Point mutations* are changes in individual genes.

Polyploidy is a condition in which cells develop extra sets of chromosomes: 3N, 4N. These extra sets of chromosomes change the characteristics of organisms. Usually polyploidy occurs in plants. Breeders may treat special plants with a chemical such as colchicine which prevents the division of the cell after the nucleus has divided. Polyploid fruits and flowers are quite large.

SEX DETERMINATION

In human beings, there are 22 pairs of *autosomes*, chromosomes that affect all characteristics except sex determination. One pair of chromosomes determines the sex of an individual. In females, the sex chromosomes are designated as XX. In males, the sex chromosomes are XY. Certain disorders are sex-linked, usually passed from mother to son by a defective gene on the X chromosome; red-green color blindness is one such sex-linked trait that is found more frequently in males and hardly at all in females. Hemophilia is another sex-linked trait that affects males with greater frequency than females.

MULTIPLE ALLELES

Alleles are two or more genes that have the same positions on homologous chromosomes. Alleles are separated from each other during meiosis and come together again at fertilization when homologous alleles are paired, one from the sperm cell and one from the egg cell. Two or more alleles determine a trait.

The inheritance of some characteristics cannot be explained by the action of a single pair of alleles. It has been shown experimentally that multiple alleles determine certain traits. However, within each cell no more than two alleles are present.

The inheritance of the ABO blood group in humans is an example of the existence of multiple alleles: I^A, I^B, and i. In this example, alleles I^A and I^B are codominant with each other and i is recessive to both I^A and I^B. Table 14.1 shows blood types and possible genotype.

TABLE 14.1. Blood Genotypes

Blood Type	Genotype
A	$I^A I^A$ or $I^A i$
B	$I^B I^B$ or $I^B i$
AB	$I^A I^B$
O	ii

Modern Genetics

THE ROLE OF NUCLEIC ACIDS

DNA is an important part of the chromosome structure of all cells. DNA is a nucleic acid as is RNA. The unit of structure and function in the nucleic acid is called a *nucleotide*. A nucleotide is composed of a *phosphate group*, a five-carbon sugar, and a protein base. If the five-carbon sugar is *ribose*, the nucleic acid is *ribonucleic acid* (RNA). If the five-carbon sugar is deoxyribose, then the nucleic acid is *deoxyribose nucleic acid* (DNA). Fig. 14.1 is a diagrammatic presentation of a nucleotide.

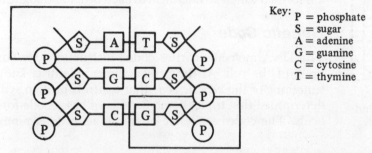

Key: P = phosphate
S = sugar
A = adenine
G = guanine
C = cytosine
T = thymine

Fig. 14.1 Nucleotides are the units on which DNA molecules are built.

The protein bases in nucleic acids are ring compounds. Those bases with single rings are *pyrimidines*. Bases with double rings are *purines*. The pyrimidines in nucleic acid are *thymine, cytosine,* and *uracil*. The purines are *adenine* and *guanine*. The four bases that make up the DNA molecule are adenine (A), guanine (G), thymine (T) and cytosine (C). The four bases that make up the RNA molecule are adenine (A), guanine (G), cytosine (C) and uracil (U).

In the early 1950's, James Watson, an American, and Francis Crick, an English investigator, unraveled the structure of DNA. The Watson-Crick model of DNA, as it has come to be known, indicates that the DNA molecule is shaped like a double helix. It consists of two long chains of nucleotides turned around each other in the shape of a double spiral. Fig. 14.2 shows the DNA molecule as a double helix. Notice that this diagram resembles a step ladder that has curved sides and straight rungs. The sides of the DNA molecule consist of alternate phosphate-sugar groups. The rungs of the "ladder" consist of protein bases joined by weak hydrogen bonds. It is known that adenine and thymine link together while guanine and cytosine link together (Fig. 14.3).

The RNA molecule is usually single stranded and much smaller than DNA. RNA is synthesized in the nucleus and functions in the cytoplasm where it controls the synthesis of proteins by the ribosomes.

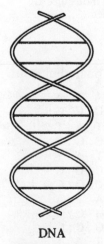

DNA

Fig. 14.2 DNA: double helix

HOW DNA FUNCTIONS

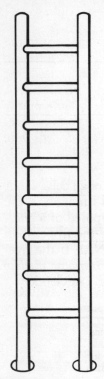

Deoxyribonucleic acid is able to function as genetic material because it has unusual characteristics. It is stable, can make more of itself, and can control the production of enzymes. DNA is stable as demonstrated by its ability to pass hereditary traits from one generation to another unchanged. However, DNA does change occasionally. This accounts for mutations.

DNA Replication

DNA can make more of itself in a process called *replication*. Replication refers to a duplication of molecules. Prior to the onset of cell division (mitosis and meiosis) DNA molecules replicate in a way that is at the same time both simple and precise. The double-stranded helix unwinds, forming a structure that resembles a straight-sided ladder with horizontal rungs. The rungs of the ladder are formed by the joining of a pyrimidine molecule with a purine through a weak hydrogen bond. The bond breaks and the two strands "unzip" between the base pairs. Complementary nucleotide chains are joined to the free base ends in the unzipped strands. Fig. 14.4 illustrates replication of DNA. As the result of the incorporation of free nucleotides into the unzipped strands, two new molecules of DNA are formed which are identical to each other and to the original molecule.

Fig. 14.3
Straight-sided
DNA

Genetic Code

The DNA molecule carries coded instructions for controlling all functions of the cell. At the present time, scientists know most about the functions of the *genetic code* that controls protein synthesis. They have determined that triplet combinations of bases code for each of 20 amino acids. The coded sequence of amino acids determines the formation of

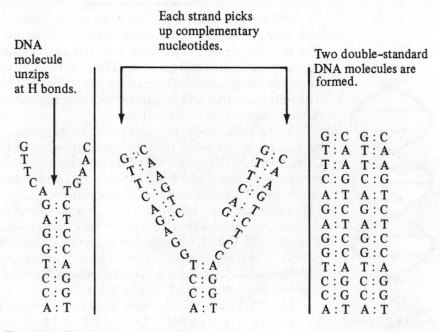

Fig. 14.4 The replication of DNA

different types of proteins. The code for proteins is present in messenger RNA molecules which are complementary to DNA molecules. For example: let us suppose that a portion of a DNA molecule carries a code such as AAC GGC AAA TTT. Its mRNA complement would be as follows: UUG CCG UUU AAA.

The mRNA, carrying this code, now moves from the nucleus to the cytoplasm. The mRNA attaches itself to several ribosomes, each having its own ribosomal RNA. Specific transfer RNA (tRNA) molecules bring to the ribosomes their own kind of activated amino acids. Transfer RNA molecules that fit the active sites of mRNA's on the ribosomes temporarily attach to them. As a result, amino acids are lined up in the proper sequence (Fig. 14.5). Note that the RNA code is a *triplet code* with one triplet, or *codon*, made up of three bases coding for a specific amino acid.

One Gene–One Polypeptide Hypothesis

In 1941 Beadle and Tatum used the red bread mold *Neurospora crassa* to find out how genes influence the synthesis of enzymes. As a result of their work the "one gene—one enzyme" hypothesis was formed. This hypothesis suggested that the synthesis of each enzyme in a cell was controlled by the action of a single gene. At present, it is known that a single enzyme may be composed of several polypeptides. The synthesis of each polypeptide is governed by a different gene. Hence, the new hypothesis "one gene—one polypeptide" is considered to be more accurate.

POPULATION GENETICS

A population includes all members of a species that live in a given location. Modern geneticists are concerned about the factors in populations that affect gene frequencies. All of the genes that can be inherited (heritable genes) in a population are known collectively as the *gene pool*. The Hardy-Weinberg Principle uses an algebraic equation to compute the gene frequencies in human populations. The conditions set by the Hardy-Weinberg Principle for determining the stability of a gene pool are as follows: large populations, random mating, no migration, and no mutation.

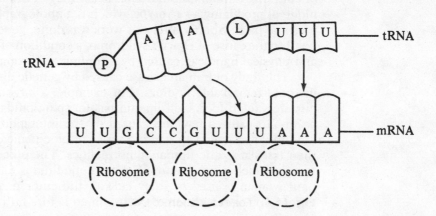

Fig. 14.5
Protein synthesis

SOME NEW DIRECTIONS IN GENETICS

Modern scientists have found it possible to transfer genetic information from one organism to another. This results in the production of *recombinant DNA*. As a result of such transfers, new genes can be introduced into an organism. Once done, the cell can synthesize the protein coded for by the newly acquired genes. Proteins that have been produced in this way are interferon, insulin and human growth hormone.

Importance to Humans

The concept of heredity has always been important to humans. The ancients living five thousand years ago in Babylonia and Assyria carried out the process of pollination to make their date palms produce fruit. Since they did not understand the mechanisms of fertilization and genetics, they turned to superstitious practices to solve their problems.

Plant and animal breeders throughout history have engaged in practices to improve their crops and their domesticated animals. Using the method of *selection* they mated organisms based on observed characteristics which were desirable to the breeder. But selection has its limits. Species cannot be improved beyond their genetic capabilities. Species improvements have been made through hybridization, in which varieties of organisms are crossed (mated) in the hope that the favorable characteristics of each show up in the offspring. For example: a cross between the Texas longhorned cattle and the Indian Brahman bull has resulted in hybrids that are resistant to infections and produce good quality meat.

HUMAN GENETICS

In 1956 the work of Joe Hin Tjio and Albert Levan provided a method of counting human chromosomes accurately. They developed the technique of producing a *karyotype*, which is a photograph of matched chromosome pairs. Shortly after this work was done, geneticists were able to point to the cause of Down's syndrome, a condition of mental retardation and physical handicap caused by a tripling of chromosome number 21.

The study of human disease caused by genetic disorders has resulted in some remarkable findings. For example: a large number of human disorders (club foot, cleft lip and palate, spina bifida, and water on the brain—to name a few) are caused by the interaction of several genes. One mutated gene can cause a series of biochemical defects which are then translated into human abnormalities. The procedure of *amniocentesis* in which a small amount of amniotic fluid is removed from a pregnant woman is used to study cells of the embryo. In this way certain chromosomal defects can be determined before birth.

Chronology of Famous Names in Biology

1863	**Gregor Mendel** (Austrian)—initiated the first mathematical study of inheritance.
1869	**Fredrich Meischer** (German)—extracted nucleic acid from cells and named the substance.
1900	**Hugo deVries** (Dutch)—discovered mutations in the evening primrose.
1903	**Walter S. Sutton** (American)—discovered the mechanism of meiosis while studying the sperm cells of grasshoppers.
1908	**Sir Archibald Garron** (English)—discovered "inborn errors of metabolism".
1911	**Thomas Hunt Morgan** (American)—developed the theory of the gene. He was the first investigator to use the fruitfly for genetic research.
1914	**Robert Feulgen** (German)—found that fuchsin-red dye is specific for DNA.
1928	**Frederick Griffith** (English)—accidentally discovered that harmless diplococci could be transformed into harmful bacteria that cause pneumonia.
1941	**George Beadle** and **Edward L. Tatum** (American)—developed the "one gene-one enzyme" theory in work with *Neurospora crassa*.
1944	**Oswald T. Avery** (American)—found the transforming factor in DNA.
1949	**P. A. Levene** (American)—showed that nucleic acid could be broken down into four nitrogenous bases.
1950	**Edwin Chargaff** (American)—found that nitrogenous bases in DNA do not occur in equal proportions.
1952	**Alfred Hershey** and **Martha Chase** (English)—performed experiments with labeled viral DNA showing that it carries the complete hereditary message.
1952	**Alfred Mirsky** (American)—showed that tissue cells contain equal amounts of DNA.
1952	**James Watson** (American)—and **Francis Crick** (England)—were the first to demonstrate the double helix structure of DNA.
1952	**Norton Zinder** and **Joshua Lederberg** (American)—discovered that DNA is the genetic material.
1958	**Linus Pauling** (American)—found a difference in electrophoresis patterns between sickle cell hemoglobin and normal hemoglobin.

1959   **Vernon Ingram** (American)—showed that the difference between normal and sickle cell hemoglobin is one amino acid in 300.

1960   **Hans Gruneberg** (Indian)—discovered that a single mutant gene in rats causes a complex of disorders.

1960   **Arthur Kornberg** (American)—first to synthesize DNA *in vitro*.

1961   **Francis Crick** (English)—provided experimental evidence supporting the triplet code while working with the T₄ virus.

Words for Study

adenine	gene	purine
alleles	gene pool	pyrimidine
codominance	genetic code	recessive
codon	genotype	replication
crossing over	guanine	RNA
cytosine	intermediate inheritance	thymine
deletion	mutation	translocation
dominant	phenotype	triplet code
DNA	point mutation	uracil
duplication		

Questions for Review

PART A.   Completion. Write in the word that correctly completes each statement.

1. The first organized study of heredity was made by ..1..
2. Characteristics that can be observed are called ..2..
3. Genes on the same chromosome are said to be ..3..
4. Short wing in fruitflies is known as ..4.. wing.
5. Linkage groups are broken by ..5..
6. The chemical ..6.. induces polyploidy.
7. Two or more genes that have the same positions on homologous chromosomes are said to be ..7..
8. Pyrimidines are nucleic acids with ..8.. rings.
9. The complete name for DNA is ..9..
10. Messenger RNA is synthesized in the ..10..

PART B. **Multiple Choice.** Circle the letter of the item that correctly completes each statement.

1. If two organisms that exhibit contrasting traits are crossed, the trait that shows up in the F_1 generation is called
   (a) codominant   (c) recessive
   (b) dominant   (d) allelic

2. The trait that remains hidden in the F_1 generation is the
   (a) dominant   (c) recessive
   (b) codominant   (d) phenotype

3. The specimen used by Mendel in his work was the
   (a) fruitfly   (c) firefly
   (b) snapdragon   (d) garden pea

4. A 3:1 ratio is characteristic of an
   (a) F_1 cross   (c) F_3 cross
   (b) F_2 cross   (d) F_4 cross

5. The genotypic ratio of an F_1 cross is
   (a) 3:1   (c) 1:2:1
   (b) 8:2   (d) 1:3:1

6. *Drosophila melanogaster* is a
   (a) fruit fly   (c) bread mold
   (b) gene   (d) cross over

7. Red cattle crossed with white cattle produce a red and white hybrid. This type of inheritance is known as
   (a) mutation   (c) recessive
   (b) codominance   (d) transformation

8. A loss of a piece of chromosome is known as a
   (a) translocation   (c) deletion
   (b) transfiguration   (d) duplication

9. Homologous chromosomes intertwine during
   (a) meiosis   (c) transduction
   (b) mitosis   (d) translocation

10. Human blood groups are formed by
   (a) a single pair of genes only
   (b) multiple alleles
   (c) several pairs of genes
   (d) no genes

PART C. **Modified True-False.** If a statement is true, write "true" for your answer. If a statement is incorrect, change the underlined word to one that will make the statement true.

1. Point mutations are changes in single <u>cells</u>.

2. When a piece of broken chromosome sticks to another complete chromosome, the defect is known as <u>linkage</u>.

3. The five-carbon sugar in DNA is named <u>ribose</u>.

4. The sex chromosomes of human males are <u>XX</u>.

5. A person with genotype <u>ii</u> has blood type <u>AB</u>.

6. Thymine is always joined to <u>guanine</u>.

7. The DNA molecule is shaped like a double <u>circle</u>.

8. DNA molecules are composed of units called <u>nucleic acids</u>.

9. RNA does not contain the base <u>uracil</u>.

10. Pyrimidine molecules are joined to purines through weak <u>oxygen</u> bonds.

Answers to Questions for Review

PART A

1. Mendel
2. phenotypes
3. linked
4. vestigial
5. crossing over
6. colchicine
7. alleles
8. single
9. deoxyribonucleic acid
10. nucleus

PART B

1. b
2. c
3. d
4. b
5. c
6. a
7. b
8. c
9. a
10. b

PART C

1. genes
2. duplication
3. deoxyribose
4. XY
5. 0
6. adenine
7. helix
8. nucleotides
9. thymine
10. hydrogen

CHAPTER 15

PRINCIPLES OF EVOLUTION

Evolution concerns the orderly changes that have shaped the earth and that have modified the living species that inhabit the earth. Evolution is a fusion of biological and physical sciences that have provided supporting data which confirm the fact that over periods of time major changes have occurred in the interior of the earth and on its surface, accompanied by modifications in climate. All of the changes in the earth are classified as nonbiological or *inorganic* evolution. Changes that have taken place in living organisms are known as biological or *organic evolution*.

Evidence of Evolution

Evidence that evolution—gradual change over a period of time—has occurred in living things is provided by many sciences and includes facts from the geologic record, the study of fossils, and evidence from cell studies, biochemistry, comparative anatomy and comparative embryology.

THE GEOLOGIC RECORD

Look at Table 15.1—A Geological Time Scale. This scale is to be read from the bottom upward because it represents the age of the earth as determined by the rock layers of the earth. Geologists believe that the earth is between 4.5 and 5 billion years old. The age of the earth is measured by a process called *radioactive dating*.

TABLE 15.1. The Geologic Time Scale

Era	Period	Epoch	Millions of Years Ago	Climate and Life
CENOZOIC	Quarternary	Recent	0.01	4 ice ages; *Homo sapiens*.
		Pleistocene	2.5	Increase in herb population; first *Homo*.
	Tertiary	Pliocene	7.0	Cool; hominoid apes; first humans.
		Miocene	26.0	Forests decrease; dominance of angiosperms.
		Oligocene	38	Dominance on land by mammals, (anthropoid apes, ungulates, whales); birds; insects.
		Eocene	54	Mild to tropical weather; small horses.
		Paleocene	65	First primates and carnivores.
MESOZOIC	Cretaceous		136	Rise of angiosperms; decline of gymnosperms; extinction of dinosaurs; second great radiation of insects.
	Jurassic		190	Europe covered by ocean; last of the seed ferns; gymnosperms dominant; reptiles dominant; origin of birds; dinosaurs abundant.
	Triassic		225	Extensive arid and mountainous areas; dominance of land by gymnosperms; first dinosaurs; first mammals.
PALEOZOIC	Permian		280	Glaciers in southern hemispheres; Appalachians rising; first conifers, cycads, ginkos; expansion of reptile; decline of amphibians; extinction of trilobites.
	Carboniferous	Pennsylvanian Mississippian	345	Subtropical climate; swamps; great coal forests; lycopsids, sphenopsids, ferns, gymnosperms; reptiles evolve; amphibians dominant; first insects; fungi.
	Devonian		395	U.S. covered by oceans, Europe mountainous; primitive tracheophytes; origin of first seed plants; first liverworts; age of fishes, sharks.

TABLE 15.1. The Geologic Time Scale (*Continued*)

Era	Period	Epoch	Millions of Years Ago	Climate and Life
PALEOZOIC (*cont'd*)	Silurian		430	Rise of mountains in Europe; continents flat; first vascular plants; arthropods appear on land.
	Ordovician		500	Mild climate; seas cover continents; plants invade land; marine algae abundant; jawless fishes evolve.
	Cambrian		570	Primitive marine algae; marine invertebrates in great numbers.
PRECAMBRIAN				Earth cooling; shallow seas; eukaryotes evolve; cyanobacteria; bacteria.

Scientists have determined that certain elements disintegrate by giving off radiations spontaneously and at a regular rate. Such elements are said to be *radioactive*. In the process of emitting radiations, the radioactive substance changes to something else. For example: uranium-238 changes to lead. The *half-life* of U-238, the rate at which one half of the uranium in a rock sample will change to lead, is 4.5 billion years. Uranium's rate of decay is not affected by any chemical or physical conditions. Therefore measuring the uranium-lead ratio in a sample of rock is a very reliable method for estimating the age of the rock.

Dating of the oldest rocks found on earth indicates that they are about 3 billion years old. To allow time for the original formation of the rocks, geologists add another 2 billion years to this figure, thus arriving at the 4.5 to 5 billion-year estimate of the age of the earth.

FOSSIL EVIDENCE

Fossils are the preserved remains of plants and animals. They are often found in sedimentary rock, which is formed by the gradual settling of *sediments*. The age of fossils is estimated by the use of *carbon dating*. The sample plant or animal fossil is tested for the ratio of radioactive carbon (carbon 14) to nonradioactive carbon (carbon 12). By using the rate of decay of carbon 14 to carbon 12, the age of a fossil can be determined.

The fossil records contained in the layers of sedimentary rock provide reliable evidences of change in plant and animal species. The lower down the rock layer, the older the fossil. Top layers contain more recent fossil remains of more complex species. The hard parts of animals, such as a shell or a skeleton, become fossilized in the hardened sediments of rock. Imprints, casts, or molds are other types of fossil remains, produced by an organism leaving a footprint, a track or form in the sediment.

Other types of fossil remains have also provided evidence of ancient species. *Amber* is the hardened resin of trees. Insects trapped in the sticky gum remain preserved as the resin hardens into amber. Ice has preserved some rather huge animals, such as the woolly mammoth, which was probably caught in the glaciers of Siberia. Leaf imprints have been preserved in coal while it was being formed. *Petrification*, the absorbing of mineral matter by dead plant and animals, preserves species in stone. Bogs have been the sources for preserved wood fossils. Since bacteria of decay cannot thrive in the acid environment of bogs, wood samples have been kept intact. Saber-toothed tiger fossils have been found in the La Brea tar pits in Los Angeles.

EVIDENCE FROM CELL STUDIES AND BIOCHEMISTRY

The cells of all living organisms have comparable structures that function in similar ways. All eukaryotic cells have a cell membrane, a nucleus (except mature red blood cells), cytoplasm with energy-producing mitochondria, and ribosomes where proteins are synthesized. The fact that the cells of all living things have similar structures that perform the same tasks indicates that there is evolutionary unity in all living things.

On the molecular level, there is similarity in genetic material of cells. Similar genes direct the formation of similar cell structures and similar proteins. For example: insulin produced in the hog pancreas is so similar to human insulin that hog pancreas is the source of insulin that is prescribed for diabetics. This means that the human and the hog have some very similar DNA molecules. It is not unusual for animals of various species to synthesize some proteins of similar natures. Interestingly enough, the primates (including humans and apes) and the guinea pigs show an unusual kind of biochemical relationship. These are the only vertebrate animals that cannot synthesize vitamin C from carbohydrates.

EVIDENCE FROM COMPARATIVE ANATOMY

A comparative study of the bone structures and body systems of animals from the various phyla reveals a great deal of similarity. A comparative study of the skeletal systems of vertebrates shows that many of the bones are very much alike. Much of our evidence for evolution comes from a study of *homologous* structures. Homologous structures are bones that look alike and have the same evolutionary origin although they may be used for different purposes. Fig. 15.1 shows the bones in the forelimbs of vertebrates. Notice the similarity of structure of these bones. The flipper of a whale, the arm of a human and the wing of a bird are *homologous* structures having the same evolutionary origin and maintaining similarity of structure.

A comparison of the digestive, nervous and circulatory systems of vertebrates indicates similar evolutionary origin. Biologists believe that there must have been a common ancestor from which the vertebrate line descended. Fig. 15.2 shows the brains of some common vertebrate species. Notice their similarities and differences.

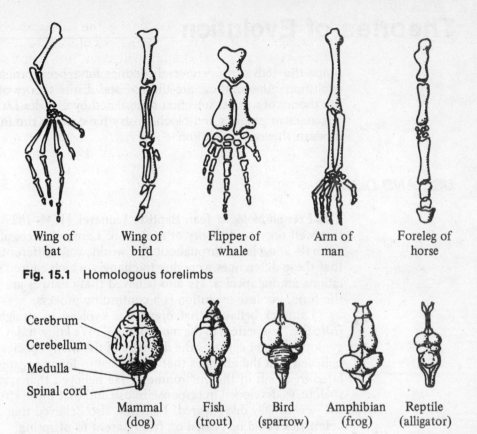

Fig. 15.1   Homologous forelimbs

Wing of bat   Wing of bird   Flipper of whale   Arm of man   Foreleg of horse

Cerebrum
Cerebellum
Medulla
Spinal cord

Mammal (dog)   Fish (trout)   Bird (sparrow)   Amphibian (frog)   Reptile (alligator)

Fig. 15.2   A comparison of vertebrate brains

EVIDENCE FROM COMPARATIVE EMBRYOLOGY

Embryology is the study of developing forms or embryos. A comparative study of the embryos of seemingly unrelated vertebrates indicates similar evolutionary origins. Fig. 15.3 shows some of these embryos. Notice the marked similarity in structure. As the embryonic development continues, the distinctive traits of each species begin to take form.

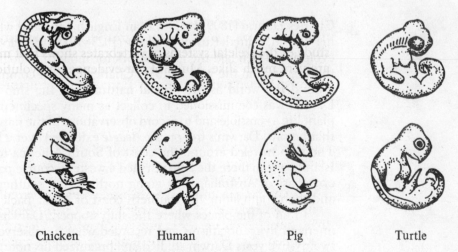

Chicken   Human   Pig   Turtle

Fig. 15.3   A comparison of vertebrate embryos

Theories of Evolution

Since the 18th century several theories have been proposed to explain evolution. Among these are the use and disuse theory of Lamarck and the theory of natural selection formulated by Charles Darwin. Recently advances in genetics and biochemistry have led to the formulation of a modern theory of evolution.

USE AND DISUSE

The French biologist Jean Baptiste Lamarck (1744–1829) presented the first well organized theory of evolution. Lamarck recognized that living animals and plants throughout the world were different. He proposed that these differences were due to changes which caused significant variations among species. He also believed that changes are going on all of the time, because evolution is a continuing process.

Lamarck believed that organisms evolved in straight line fashion from the "less perfect to the more perfect". We know today that branching evolution better explains the diversity of the many species. Lamarck also believed that the changes that take place in living organisms are stimulated by needs in the environment. He believed that systems and body structures developed in response to use and that those structures in "disuse" eventually disappeared. Lamarck also believed that acquired characteristics could be passed on from parent to offspring.

German biologist August Weismann (1834–1914) used a rather dramatic experiment to disprove Lamarck's theory of inheritance of acquired characteristics. Weismann cut off the tails of mice and mated them. He carried out this process for 25 generations of mice. In each he found that mice grew tails just as long as the tails of those mice in the first generation. Weismann proved that cutting off the tails of mice (acquired characteristic) did not alter the "germplasm" of future generations.

NATURAL SELECTION

Charles Darwin (1809–1882) was an English naturalist who together with his cousin, Alfred Russel Wallace (1823–1915), developed a theory of evolution which laid the groundwork for modern biological thinking. At the age of 22 in the year of 1831, Charles Darwin commenced on a trip around the world as the official naturalist of the ship *H.M.S. Beagle.* Darwin was commissioned to collect as many specimens of animal and plant life as possible and to record observations of the natural phenomena that he saw. Darwin's trip on the *Beagle* extended over a five-year period. The ship traveled around the coast of South America to the Galápagos Islands. From there the ship pursued a westerly course passing the southern coast of Australia, then going north to the southern coast of Asia and then south along the southern coast of Africa back to England.

In all of the places where the ship stopped, Darwin collected specimens of living organisms and recorded what he observed. For the next twenty-five years Darwin studied and organized his notes. In 1859 he put together his thoughts in a published work titled *On the Origin of Species by Means of Natural Selection, or the Preservation of Favored Races in the Struggle for Life.* In this book, Darwin's theory of natural selection was clearly set forth. Today this work is referred to by a shorter title, *The Origin of Species.*

Darwin's theory of natural selection can be summed up thusly: large numbers of new plants and animals are produced by nature. Many of these do not survive because nature "weeds out" weak and feeble organisms by killing off those that cannot adapt to changing environmental conditions. Only the strongest and most efficient survive and produce progeny. Specific tenets of the Darwin-Wallace theory of evolution follow.

Overproduction

The theory of natural selection cites the fact that every organism produces more gametes and/or organisms that can possibly survive. If every gamete produced by a given species united in fertilization and developed into offspring, the world would become so overcrowded in a short period of time that there would be no room for successive generations. This does not happen. There is a balance that is maintained in the reproduction of all species and therefore natural populations remain fairly stable, unless upset by a change in conditions.

Competition

As stated above, not all the offspring (and gametes) survive. There is competition for life among organisms: competition for food, room and space. Therefore there is a *struggle for existence* in which some organisms die and the more hardy survive.

Survival of the Fittest

Some organisms are better able to compete for survival than others. The differences that exist between organisms of the same species making one more fit to survive than another can be explained in terms of *variations.* Variations exist in every species and in every trait in members of the species. Therefore some organisms can compete more successfully for the available food or space in which to grow or can elude their enemies better. These variations are said to add survival value to an organism. Survival value traits are passed on to the offspring by those individuals that live long enough to reproduce. As weaker individuals are weeded out of the species, those individuals that remain do so because they are best adapted to live in their environment. As time goes on the special adaptations for survival are perpetuated and new species evolve from a common ancestral species.

The environment is the selecting agent in natural selection because it determines what variations are satisfactory for survival and which are not. Any change in an environment can affect the usefulness of a given variation. A change in the environment can render a once useful variation useless. In this way the direction of evolutionary development can change.

MODERN EVOLUTIONARY THEORY

The major weakness in Darwin's theory of natural selection is that he did not explain the source, or genetic basis, for variations. He did not distinguish between variations that are hereditary and those that are non-hereditary. He made the assumption that all variations that have survival value are passed on to the progeny. Like Lamarck, Darwin believed in the inheritance of acquired characteristics.

We now know that there are several ways in which variation is produced within a species.

Causes of Variation

Hugo De Vries (1848–1935), a Dutch botanist, explained variations in terms of *mutations*. His study of 50,000 plants belonging to the evening primrose species enabled him to identify changes in leaf shape and texture and in plant height that were passed on from parent plant to offspring. In 1901 De Vries offered his mutation theory to explain organic evolution. He did not know how mutations come about or where they occur. Today, we know that mutations are changes in genes that can come about spontaneously or can be induced by some mutagenic agent.

Spontaneous mutation rates are very low. In *E. coli* one cell in every 10^9 will mutate from streptomycin sensitivity to streptomycin resistance. In human beings the rate of mutation in Huntington's disease occurs in 5 out of every 10^6 gametes. It is a known fact that different genes have different mutation rates. This is probably due to the different chemical composition of genes or perhaps to the different places that they occupy on chromosomes. Mutations alone do not effect major changes in the frequencies of alleles.

An important cause of variation within species is genetic recombination that results from sexual reproduction wherein the genes of two individuals are sorted out and recombined into a new combination, producing new traits—and thus variation.

Gene flow is another agent of evolution, responsible for the development of variations. It is the movement of new genes into a population. Gene flow often acts against the effects of natural selection. *Genetic drift* is a change in a gene pool that takes place in a population as a result of chance. Let us suppose that a mutation takes place in a gene of one person. If that person does not reproduce, the gene is lost to the population. Sometimes a small population breaks off from a larger one. Within that small population is a mutant gene. Because now that the mating within the small population is very close, the frequencies of the mutant gene will increase. In the Amish population where there is very little (or no) outbreeding, there is an increase in the homozygosity of the genes in the gene pool. This is seen in the high frequencies of genetic dwarfism and polydactyly (six fingers). The isolated smaller population has a different gene frequency than the larger population from which it came. This is known as the *Founder Principle*.

Genetic drift and the random mutations that increase or decrease as the result of genetic drift are known as nonDarwinian evolution.

Time Frame

Currently, it is believed that there is a time frame for evolution. The concept of *gradualism* supports the idea that evolutionary change is slow, gradual and continuous. The concept of *punctuated equilibrium* sets forth the idea that species have long periods of stability, lasting for four or five million years, and then change as the result of some geological or other environmental change.

The Origin of Life

Theories of evolution are concerned not only with how living organisms have changed through time but also with how life began—how living organisms first evolved on earth. In the 1920's the Russian scientist A. I. Oparin began investigations into how life could have evolved from inorganic compounds under the conditions of early earth. He found that he could produce coacervate droplets that had the ability to incorporate simple enzymes in their structure. This initial work opened the door to many studies of how life could have begun.

In the 1950's Stanley Miller set up an experiment in which he duplicated the chemical conditions and the temperature of the early seas. Into sterile water he put some methane gas, hydrogen and ammonia. He sealed off the system so that nothing could leave or enter it. The water was heated to a temperature similar to that of the early seas. The water and gas mixture were subjected to electric sparks. Miller let the experiment run for about a week. At the end of that time, he tested the water and found present in it organic compounds that were not there before. Miller's experiments help to explain how organic matter appeared in the early seas.

In 1963, Carl Sagan, duplicating Miller's experiment, was able to produce ATP (adenosine triphosphate) from inorganic matter. ATP molecules are essential to all living cells and function as energy storage molecules.

Following the lead of Stanley Miller, other investigators demonstrated how organic molecules could have been formed in the early seas. Melvin Calvin demonstrated the polymerization of complex molecules. Sidney Fox produced microspheres, long chain peptides surrounded by something resembling a membrane. All of these experiments have contributed to the theory of the origin of life.

Scientists believe that the first organisms were heterotrophs that obtained nutrition from the organic molecules in the "hot, thin soup", an expression used to describe the early oceans. As oxygen began to increase in the atmosphere, conditions changed for the heterotrophs. Those heterotrophs that could not adjust to the changing atmospheric conditions were destroyed. Eventually, organisms that could make their own organic compounds from inorganic materials in the presence of sunlight began to evolve. These early autotrophs were the blue-green algae.

The blue-green algae increased the oxygen content of the air—remember oxygen is a byproduct of photosynthesis—and this further threatened the continued existence of the heterotrophs. As a result of the high concentration of oxygen in the atmosphere, an ozone layer developed which further diminished the organic compounds available to the groups of existing heterotrophs. From this group there evolved heterotrophs that could utilize oxygen in cellular respiration during which energy was released from organic molecules taken in as nutrients. Thus the carbohydrates produced by the autotrophs and the oxygen in the air supplied the newly evolved heterotrophs with the nutrients necessary for survival. This description is known as the *heterotroph hypothesis*. It is an attempt to explain the origin of autotrophic and heterotrophic cells.

Evolution of Humans _____

The evolution of *Homo sapiens* has always been of interest to modern humans. People like to know their origins. In 1924 the first specimen of *Australopithecus africanus* was discovered in South Africa. It is estimated that this prehuman form lived about 2 or 3 million years ago. Studies of the several australopithecine fossils found indicate that this species may well have been the forerunner of humans. The brain case indicates that the brain was within the size range of modern apes. The teeth were more human-like than those of modern apes. They were arranged in rows without gaps between the canines and the premolars, and the canine teeth did not stick out beyond the adjacent teeth. The bones of the pelvis show that the australophithecines were bipedal, walking on two feet. Because not all of the australopithecine fossils are alike, anthropologists believe that two species of man-ape lived at about the same time: *Australopithecus africanus* and *Australopithecus robustus*. Very little is known about the life patterns of these two species.

Fossil finds indicate that another species of prehuman lived on earth probably for as long as 500,000 to 1,000,000 years. *Homo erectus*, as this species is called, lived on various continents of the earth. The brain case is smaller than that of modern man, but the teeth are larger. The fossil teeth show traces of having had an enamel collar, a primitive trait in apes and in man. The species became extinct (Fig. 15.4).

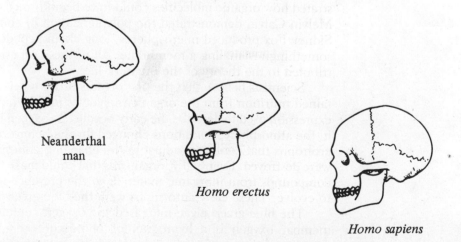

Neanderthal
man

Homo erectus

Homo sapiens

Fig. 15.4   A comparison of skulls of three human species

In 1856 the fossil remnants of a human species called *Homo sapiens neanderthales* were found in a cave in Germany. It is believed that Neanderthal man appeared on earth about 125,000 years ago. He lived through two glacial periods and then became extinct between 50,000 and 15,000 years ago. It is commonly accepted theory that over thousands of years, modern humans gradually replaced the Neanderthal grade of man, until there remained only one species of human being—*Homo sapiens* (Fig. 15.5).

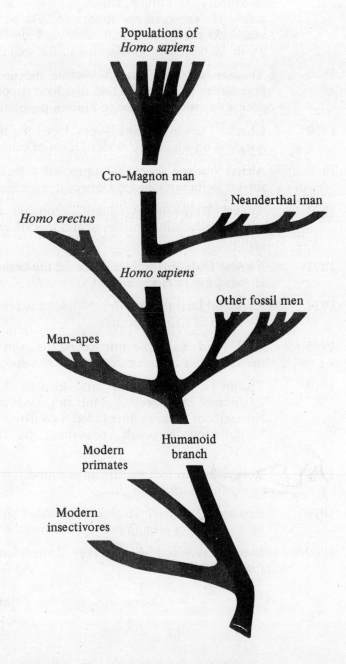

Fig. 15.5   The tree of human evolution

Chronology of Famous Names in Biology

1747   **Comte de Buffon** (French)—was the first scientist to state that living things can change.

1771   **Erasmus Darwin** (English)—was an early advocate of the idea of evolution.

1784   **Jean Baptiste Lamarck** (French)—presented the first organized theory of evolution, although his ideas are not acceptable today. He proposed the theory of "use and disuse" which suggested that organs that were well developed from use would be passed on to offspring in the well developed stage.

1800   **Thomas Malthus** (English)—wrote the notable "Essay on Population" in which he said that food supplies cannot keep pace with rapid increases in human population.

1830   **Charles Darwin** (English)—developed the theory of natural selection on which the modern ideas of evolution are based.

1830   **Alfred Wallace** (English)—proposed a theory of evolution similar to that of Charles Darwin's.

1874   **August Weismann** (German)—proposed that germplasm carries the hereditary material and this affects the successive generations of progeny.

1891   **Eugene Dubois** (Javanese)—found the braincase of the first known fossil of *Homo erectus*.

1924   **Raymond Dart** (South Africa)—identified the first recovered specimen of *Australopithecus*.

1935   **E. B. Ford** (English)—introduced the concept of industrial melanism to explain change in color of species of moths.

1940   **Trofim D. Lysenko** (Russian)—tried to change genes and inheritance by subjecting plants to periods of heat and cold. His methods in agriculture failed. Lysenko is remembered for his fallacious proposals concerning the "vernalization of wheat"

1947   **Robert Broom** (South African)—found fossil bones of *Australopithecus africanus*.

1950   **Bernard Kettlewell** (English)—provided an excellent investigative study of evolutionary selection in the peppered moth.

1960   **Louis Leakey** and **Mary Leakey** (Tanzanian)—found fossils that predate humans.

Words for Study

amber	heterotroph hypothesis
evolution	homologous
fossil	natural selection
founder principle	punctuated equilibrium
gene flow	radioactive dating
genetic drift	petrification
gradualism	

Questions for Review

PART A. Completion. Write in the word that correctly completes each statement.

1. Changes that have taken place in the structure and energy of the earth are classified as ..1.. evolution.
2. Elements that emit atomic particles are said to be ..2..
3. The kind of rock in which fossils are found is ..3.. rock.
4. Cells of all living things have ..4.. structures.
5. The flipper of a whale and the wing of a bird are ..5.. structures having the same evolutionary origin.
6. The theory of "use and disuse" was proposed by ..6..
7. The theory of natural selection was proposed by ..7..
8. Hugo De Vries explained variations in terms of ..8..
9. Loss of genes from a population is known as ..9.. (2 words).
10. The experiments of ..10.. helped to explain how organic molecules appeared in the early seas.

PART B. Multiple Choice. Circle the letter of the item that correctly completes each statement.

1. *Australopithecus africanus* is considered to have been a
   (a) man-ape
   (b) man
   (c) ape
   (d) old world monkey

2. The fossils of which of the following have been found all over the world?
   (a) *Australopithecus*
   (b) Neanderthal
   (c) *Homo erectus*
   (d) *Homo habilis*

3. The early autotrophs were probably
   (a) bacteria
   (b) viruses
   (c) amoeba
   (d) blue-green algae

4. The first living cells were probably
   (a) autotrophs (c) parasites
   (b) heterotrophs (d) symbionts

5. The modern theory of the origin of life was initially developed by
   (a) Miller (c) Sagan
   (b) Fox (d) Oparin

6. The founder principle explains
   (a) the origin of life
   (b) abnormal gene frequencies in a small population
   (c) the methods by which genes leave a population
   (d) finding and securing of lost genes

7. Genes are sorted out and recombined by the process of
   (a) genetic drift (c) sexual reproduction
   (b) mutation (d) vegetative reproduction

8. The evening primrose was the experimental specimen used in the study of
   (a) genetic drift (c) gene flow
   (b) mutation (d) vegetative propagation

9. Darwin's theory of evolution is known as
   (a) natural selection (c) survival of the fittest
   (b) struggle for existence (d) competition

10. The "continuity of the germplasm" was demonstrated in an experiment using mice by the investigator named
    (a) Miller (c) Fox
    (b) Lamarck (d) Weismann

PART C. Modified True False. If a statement is true, write "true" for your answer. If a statement is incorrect, change the underlined word to one that will make the statement true.

1. The age of the earth is measured by a process called <u>reactive</u> dating.

2. Uranium-238 ultimately turns into <u>boron</u>.

3. It is estimated that the oldest rocks on earth are about <u>5</u> billion years old.

4. Fossil plants and animals are dated by measuring the <u>hydrogen</u> content.

5. Fossil insects are most likely to be found in <u>ambergris</u>.

6. The absorbing of mineral matter by dead plant and animal bodies which turns them into stone is known as <u>petrification</u>.

7. Human beings <u>can</u> synthesize vitamin C from carbohydrates.

8. The study of developing forms is known as <u>anatomy</u>.

9. The *H.M.S. Beagle* is most closely associated with <u>Wallace</u>.

10. The concept of punctuated equilibrium helps to explain the time frame of <u>reproduction</u>.

Answers to Questions for Review

PART A

1. inorganic
2. radioactive
3. sedimentary
4. similar
5. homologous

6. Lamarck
7. Darwin
8. mutations
9. genetic drift
10. Miller

PART B

1. a
2. c
3. d
4. b

5. d
6. b
7. c

8. b
9. a
10. d

PART C

1. radioactive
2. lead
3. 3
4. carbon
5. amber

6. true
7. cannot
8. embryology
9. Darwin
10. evolution

CHAPTER 16

ECOLOGY

Ecology is the science that studies the interrelationships between living species and their physical environment. The word "ecology" was coined in 1869 by the German zoologist Ernst Haeckel to emphasize the importance of the environment in which living things function. The environment includes living or *biotic* factors and nonliving factors referred to as *abiotic* factors.

The abiotic factors consist of physical and chemical conditions that affect the ability of a given species to live and reproduce in a particular place. Included in the abiotic factors are temperature, light, water, oxygen, pH (acid-base balance) of soil, type of substrate, and the availability of minerals. Certain kinds of plants and animals will flourish in a natural community if the conditions are present that permit their survival. Species interact to influence the survival of one another. One important principle of ecology is that no living organism is independent of other organisms or of the physical environment, if they share the same community.

The Concept of the Ecosystem

Certain terms are used in ecology to provide a consistent description of conditions and events. A *population* refers to all of the members of a given species that live in a particular location. For example: a beech-maple forest will contain a population of maple trees, a population of beech trees, a population of deer, and populations of other species of plants and animals. All of the plant and animal populations living and interacting in a given environment are known as a *community*.

The living community and the nonliving environment work together in a cooperative ecological system known as an *ecosystem*. An ecosystem has no size requirement or set boundaries. A forest, a pond and a field are examples of ecosystems. So is an unused city lot, a small aquarium, the lawn in front of a residential dwelling, or a crack in a sidewalk. All of these examples reflect areas where interaction is taking place between living organisms and the nonliving environment.

Modern ecologists think of the ecosystem in terms of its interacting forces or components. One such component is the physical environment. This includes the air, which is made up of 21 percent oxygen, 78 percent nitrogen, .03 percent carbon dioxide, and the remainder inert gases. The soil is the source of minerals that supply plants with compounds of nitrogen, zinc, calcium, phosphorus and other minerals.

The green plants in the ecosystem are another component. These are the *producers*, so-called because they are able to make their own food using the inorganic materials of carbon dioxide and water and minerals from the soil. Directly dependent upon the producers are the *primary consumers—herbivores*, or plant-eaters. Herbivores come in all sizes: crickets, leaf cutters, deer and cattle. The *carnivores*, or flesh-eaters, such as snakes, frogs, hawks, and coyotes are *secondary consumers* because they feed on the herbivores. The *tertiary consumers* are those that feed on the smaller carnivores and herbivores as well. There are also scavengers in the ecosystem. Earthworms and ants feed on particles of dead organic matter that have decayed in the soil. Vultures eat the bodies of dead animals.

The *decomposers* form another important part of ecosystems. Bacteria and fungi are organisms that break down dead organic matter and release from it organic compounds and minerals that are returned to the soil. Many of the materials returned to the soil are used by the producers in the process of food-making. Without the work of the decomposers the remains of dead plants and animals would pile up, not only occupying space needed by living organisms, but also keeping trapped within their dead bodies valuable minerals and compounds.

The structure of the ecosystem remains the same whether its location is on land or in water. A marine ecosystem illustrates this fact. The autotrophs are microscopic plants called *phytoplankton* which float on the top of the water. The phytoplankton is composed of billions of single-celled green algae which produce a vast quantity of food by means of photosynthesis just as the land plants do. Living in close association with the phytoplankton are heterotrophs such as protozoa and brine shrimp. Since these species feed on the algae, they are also herbivores. Other types of consumers are dependent upon the algae also. Among these are the secondary consumers, small fish that constitute the carnivores that feed on the herbivores. Then there are the larger fish that feed on the smaller species of fish. Dwelling on the bottom sediments are the scavenger worms and snails that feed on the dead plant and animal bodies that fall to the bottom. Living among the sediments are the decomposers—the fungi and bacteria—that break apart plant and animal remains.

THE ECOLOGICAL NICHE

An important concept of ecology is that of the *niche*. An ecological niche is a feeding pattern exhibited by species that compose a community. A niche is a feeding way-of-life in relationship to other organisms. For example: small woodpeckers and nuthatches feed on grubs that are present in the crevices of trees. Although the woodpecker and the nuthatch feed on grubs present in the same tree, these species are not in competition with each other. The woodpecker searches for its food in the crevices at the bottom of the tree and works its way upward. The nuthatch functions best at the top of the tree and works its way downward. Both of these species may live close together, may feed on similar organisms on the same tree, but select the grubs from different locations on the tree. Therefore, it can be said that the woodpecker and the nuthatch occupy different ecological niches.

When two species live in the same place and feed in the same way, using the same food at the same time, *competition* results. The species that has special adaptations for reaching the food first or has greater reproductive potential will be the survivor. The other species will be eliminated. The process by which elimination establishes one species per niche in a particular habitat is known as the *competition-exclusion principle*.

Energy Flow in an Ecosystem

The source of all energy in an ecosystem is the sun. Autotrophs are able to capture just a small portion of the sun's energy and use it to make food enough for all of the living organisms in the ecosystem. The green plant is able to store energy temporarily in ATP molecules and in the nutrients that it makes. Energy is then transferred from green plants into the animal body where it is used to power the vital functions necessary for life. Once energy is used to do work it is converted into heat which then escapes the body and radiates into the atmosphere. The cycles of photosynthesis (energy trapping and conversion) and respiration (energy release and use) must be repeated *ad infinitum* if the ecosystem is to continue.

ENERGY AND FOOD CHAINS

The flow of energy through an ecosystem can be studied by way of *food chains*, which show how energy is transferred from one organism to another through feeding patterns. An example of a food chain on a cultivated field might be as follows:

Lettuce → Rabbit → Snake → Hawk

Each stage in the food chain represents a feeding or *trophic level*. Lettuce is the producer, the green plant that provides the food which supports the ecosystem. The rabbit is the primary consumer and represents the second trophic level. The carnivorous snake feeds upon the rabbit and thus represents the third trophic level, a secondary consumer. This food chain ends with the hawk, a tertiary consumer that occupies the smallest trophic level in terms of energy. Every food chain begins with an autotroph and ends with a carnivore that is not eaten by a larger animal. When carnivores die, their bodies usually serve as food for scavengers. The bacteria and fungi of decay decompose the remains not devoured by the scavengers.

The flow of energy in a food chain is in a straight line pattern. Most of the energy is concentrated in the level of the producer. At each succeeding level the energy is decreased. However, the feeding relationships among organisms in an ecosystem are not usually this simple, and, in actuality, are more complex.

ENERGY AND FOOD WEBS

Let us look again at a simple food chain.

Lettuce → Rabbit → Snake → Hawk

Suppose all of the rabbits disappear from the cultivated field. Will the snakes die of starvation? The answer to this question can be seen in Fig. 16.1, which illustrates a food web. Notice that the snake can feed upon a shrew, mouse, or owl. In other words, other primary consumers (herbivores) serve well as food for snakes. Fig. 16.1 shows that there are several alternative energy pathways in a *food web*. It is the alternative pathways that enable an ecosystem to keep its stability. One species does not eradicate another in the quest for food.

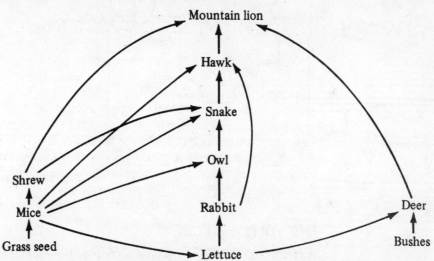

Fig. 16.1 A food web

The relationships between predators and prey help to keep an ecological community stable. The predators help to keep the populations upon which they feed in check. For example: if rabbits were allowed to reproduce without their numbers being thinned by natural predators, the population of rabbits would become overwhelming. The burgeoning rabbit population would devour the available vegetation and then the species would experience starvation and death. Such imbalances in nature do occur when humans upset the natural stability of an ecosystem. Past experience has shown that ecosystems are upset when new species are introduced into an area where there are no predators or natural enemies of the species. Fig. 16.2 shows a food pyramid, another way of illustrating energy flow in an ecosystem. Notice that the autotrophs at the base of the pyramid support all of the heterotrophs (consumers) that exist at each nutritional level and that there is a decrease of available energy at each nutritional level.

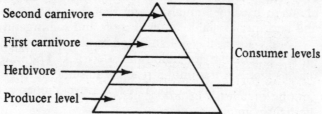

Fig. 16.2 Food pyramid

Biogeochemical Cycles _____

Certain compounds cycle through the abiotic portion and the biotic communities of ecosystems. These compounds contain elements that are necessary to the biochemical processes that are carried out in living cells. Among these elements are carbon, hydrogen, oxygen and nitrogen. In elemental form, they are useless to cells and must be combined in chemical compounds. Let us trace the pathways of some of the vital compounds from the earth to living organisms to the atmosphere and back to earth. This cycle of events is best described by the term *biogeochemical cycles* (Fig. 16.3).

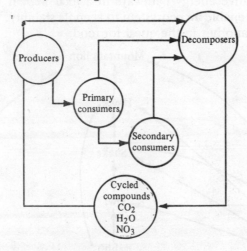

Fig. 16.3 Biogeochemical cycles

THE WATER CYCLE

Water is the source of hydrogen, one of the elements necessary for the synthesizing of carbohydrate molecules by green plants. Water is necessary as the dissolving medium for substances that cross cell membranes and enter cells. Without water, there can be no life.

There are three ways in which water vapor enters the atmosphere. Water evaporates from land surfaces and from the surfaces of all bodies of water. Water vapor enters the air as a waste product of respiration of animals and plants. For example: every time you exhale water vapor is released into the air. Great amounts of water are lost from plants through the openings in the leaves; this water loss due to evaporation is called *transpiration*.

Water vapor in the air is carrried to high altitudes where it is cooled and forms clouds by condensation. Eventually, clouds fall to the earth in the form of precipitation: rain, snow, or sleet. Most of the precipitation returns to the oceans, lakes or streams and less than 1 percent of it falls on land. Of the water that does fall on land, about 25 percent of it will evaporate from the various land surfaces before it can be absorbed by plants or used by animals. Water that does not evaporate enters the soil and becomes available to plant roots and soil organisms.

Soil water that is not absorbed by plants seeps down into the ground until it reaches an impervious layer of rock. The water moves along this rock as *groundwater* until it reaches an outlet into a larger body of water such as a lake or an ocean. The water cycle repeats. Fig. 16.4 illustrates the water cycle.

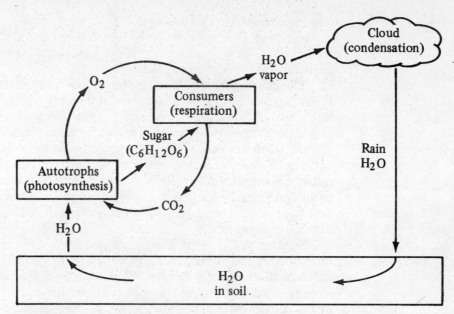

Fig. 16.4 Water cycle

CARBON DIOXIDE/OXYGEN CYCLE

Carbon dioxide and oxygen cycle through the abiotic and biotic components of ecosystems. Oxygen is released from plants into the atmosphere during photosynthesis. Oxygen from the air is taken in by animals and used during cellular respiration. Carbon dioxide, a waste product of respiration, is released into the air by animals and plants. Plants take in the carbon dioxide and use it during photosynthesis. The cycle repeats.

NITROGEN CYCLE

About 80 percent of the volume of the air is elemental nitrogen. Although it is an important component of proteins, nitrogen must be in a combined form before it can be used by the cells of living things.

Nitrogen becomes available to plants through the action of nitrogen-fixing bacteria that change atmospheric nitrogen into nitrates which are used by the plants for protein synthesis.

Nitrogen is also fixed by lightning. The energy produced by a discharge of lightning joins nitrogen with oxygen to form nitric oxides. These oxides combine with water vapor to form nitrous and nitric acids. When these acids fall on soil with rainwater, they are changed into nitrites and nitrates.

Denitrifying bacteria in the soil change nitrates back to atmospheric nitrogen. The cycle then repeats.

The Limiting Factor Concept

There are environmental factors such as light, temperature, and amount of rainfall that set the limit of conditions for species survival. Every species can live within a range of values for each factor in its physical environment but cannot survive beyond the low and high values called the *limits of tolerance*. For example: the eggs of the frog *Rana pipiens* exhibit a limit of tolerance for the temperature range between 0° and 30°C. The eggs of *Rana pipiens* will die in temperature ranges below or above these figures. However, at 22°C more *Rana pipiens* eggs hatch than at any other temperature. This means that 22°C is *optimum temperature* for the survival of this species. Among factors in the physical environment that set limits on growth and other physiological processes are light, temperature and amount of rainfall.

The effect of light as a limiting factor was demonstrated in 1920 when W. W. Garner and H. A. Allard showed that daylength was an important factor in the flowering of plants. The physiological response made by plants to changes in daylength is known as *photoperiodism*. Some plants (short-day) flower only if they are exposed to light for *less than* a certain amount of time each day; other plants (long-day) must have a certain minimum length of photoperiod. Researchers explain the ability of plants to measure time by the action of phytochrome, a light absorbing pigment that is associated with the cell membrane and with some of the cell's internal membranes.

Between 0°C and 40°C is the range of temperature between which most species are active. In temperatures below 0°C there is a slowing down of biochemical processes. Freezing causes spicules of ice to form in cells which causes the lysing of cells and the disruption of cell processes. Above 40°C protein of most species denatures, rendering enzymes, hormones and the structural proteins of cells useless. However, some forms of life are able to exist outside the range of 0° to 40°. The seeds of most deciduous trees must go through a period of cold, very often in temperatures far below the freezing point of water, before they can germinate. It is also a fact that certain microorganisms such as bacteria and certain species of algae can live in hot springs and geysers where the temperature may exceed 90°C.

Warm-blooded mammals that live in cold climates have developed physiological mechanisms for preserving body heat. The combination of large amounts of body fat and the ability to hibernate during the most forbidding of the cold months serves to maintain the lives of these organisms. On the other hand, cold-blooded animals have body temperatures that are quite similar to the temperature of the external environment. They too become inactive in cold weather and then require protective shade during the hot season.

Temperature is a factor that regulates the growth of plants and the activities of animals. There is no doubt that plant and animal hormones evoke physiological changes in response to the rise and fall of temperature.

Rainfall is also a limiting factor. Land-dwelling plants and animals depend upon rainfall as the source of water. Species that live successfully in deserts or in areas where there is seasonal drought have developed adaptations for water conservation. The kangaroo rat lives in humid burrows under the desert sands where, huddled together with others of its kind, it licks the condensed drops of water vapor from the hair of its burrow mates. The lungfish (*Dipnoi*) estivates under the mud in dried ponds, awaiting the seasonal downpour of rain that will fill up its habitat and restore the species to active life. The seeds of several flowering plant species of the desert will germinate only when a heavy rain wets the substrate around them; the rainfall seems to remove from the seed coat a chemical compound that inhibits germination. Similarly, species in areas of heavy rainfall have special adaptations. Some plant species in the rain forests of Brazil have developed adaptations such as aerial roots that can absorb moisture from the air rather than becoming waterlogged in too wet soil.

Ecological Succession

It was stated before that an ecosystem consists of a living community and its nonliving environment. As physical conditions change in an ecosystem, so do the plants and animals that inhabit that system. The orderly change of the biotic community in an ecosystem is known as *ecological succession*. There are two levels of succession: primary and secondary.

As the name implies, primary succession is the introduction of living species into a region that was formerly barren. Let us suppose that a rocky cliff is devoid of soil. By the processes of erosion and weathering, the rock is pulverized. Lichens that inhabit the bare rock contribute to its chemical erosion. Dead lichens contribute organic matter to the pulverized rock material. Now the physical conditions have changed just enough to allow the growth of mosses. These too add organic matter to the substrate and also contribute some water-holding ability to the newly formed soil. The moss population is followed by the grasses and then the ferns. In orderly succession there follow the low-growing bushes, the higher-growing shrubs and then trees. Each population inhabiting the region makes it better for another species.

Primary succession culminates in the existence of a stable community, which is known as a *climax community* where one or two large trees predominate. Over a period of years secondary succession may occur in which one species of tree is gradually replaced by another. For example: white pine trees may replace the gray birch trees. Limiting factors such as temperature, light, rainfall and mineral content of the soil influence the type of climax community. A climax community in a broad geographical area having one type of climate is known as a *biome*.

World Biomes

The earth is divided into several biomes.

THE TUNDRA

Vast stretches of treeless plains surrounding the Arctic Ocean where cold is the limiting factor (60°F in the summer to − 130°F in the winter) are known as the arctic tundra. Here the ground is permanently frozen a few feet below the surface and is responsible for the many lakes and bogs that characterize the region. The kinds of organisms that inhabit the tundra are those that have developed adaptations for survival in extremely cold temperatures. The plant species is composed of lichens, mosses, grasses and sedges. During the summer, the flowering herbs bloom in brilliant color for a very short time along with the dwarf willows. This seemingly meager plant life is able to support the food chains of the tundra. It feeds thousands of migrating birds and insect swarms that appear during the brief summer. Many mammals such as the musk ox, caribou, polar bears, wolves, foxes and marine mammals remain active for most of the year.

THE TAIGA

The coniferous forests of Canada survive well in the long severe winters. The spruce and fir trees predominate. The kinds of mammals that inhabit the taiga are the black bear, the wolf, lynx and squirrel.

THE DECIDUOUS FOREST

The forests of the temperate regions are dominated by broad-leaved trees that lose their leaves in the winter. Examples of the kinds of trees that compose these hardwood forests are oak and hickory, oak and chestnut, beech and maples and willows, cottonwood and sycamore. The types of animals inhabiting these forests are deer, fox, squirrel, skunk, woodchuck and raccoon.

THE DESERT

Deserts form in regions where the annual rainfall is less than 6.5 centimeters, or where rain occurs unevenly during the year and the rate of evaporation is high. The temperature changes drastically from hot days to cold nights. Plants that survive in the desert have specific adaptations for low moisture and high temperature. Shrubs, such as creosote and sagebrush, shed their short, thick leaves during dry spells and become dormant as protection against wilting. Fleshy desert plants, such as the cacti of American deserts and the euphorbias of African deserts, store water in their tissues. Cheat grass and wild flowers are annuals which grow quickly after a desert downpour, bloom, produce seed and die. Mosses, lichen and blue-green algae lie dormant on the sand and become active when moisture is present. The animals of the desert include lizards, insects, kangaroo rats and arachnids.

THE GRASSLANDS

Grasslands occur where the annual rainfall is low and irregular. Grasslands usually occupy large areas of interior continents that are sheltered from moisture-laden winds. In the United States the grasslands are known as the Great Plains; in Russia they are called the *steppes*, in South Africa, the *veldt* and in South America as the *pampas*. Grasses have adaptations for living in soil where the rainfall is low and erratic. Grazing animals are suited for life on the grasslands. At one time the American grasslands supported huge herds of antelope and bison. The settlers replaced these natural herds with cattle and sheep. Besides the domesticated animals, a number of predator species inhabit the grasslands, including coyotes, bobcats, badgers, hawks, kit foxes and owls. These feed primarily on burrowing rodents.

THE TROPICAL RAIN FOREST

The tropical rain forest is characterized by high temperatures and constant rainfall. This type of biome is found in Central and South America, in Southeast Asia and in West Africa. The trees are tall and the vegetation is so thick that the forest floor is shaded from light. The animals of the rain forest include monkeys, lizards, snakes and birds.

THE SEA

Ocean waters cover almost three fourths of the earth's surface and support the greatest abundance and diversity of organisms in the world. Averaging 3.5 to 4.5 kilometers in depth, a marine biome constitutes the thickest layer of living things in the biosphere. The dominating physical factors determine the type of living organisms that compose its communities. Temperature, light intensity, salinity, waves, tide currents and pressures are the limiting factors that set the conditions for life.

Light in the ocean extends for a depth of 180 meters; beyond this there is total darkness. The upper layers of water are known collectively as the continental shelf, a region that supports a vast array of living things. Phytoplankton live at the surface and serve as food for primary consumers such as sardines and anchovies. These herbivores serve as food for larger fish such as salmon, mackerel and tuna. Mud-burrowing animals such as clams, snails, worms and shrimp are eaten by crabs, lobsters, starfish, and a number of different kinds of fish such as cod, halibut, haddock, rays, and flounder. In the dark region of the sea, phytoplankton are absent. The small animals that live at the bottom of the sea feed on organic debris that falls from above. The temperature of the sea is fairly constant. Its salinity lowers the freezing point of water.

Conservation

The conservation of natural resources is of primary importance to humans if the land on which they live is to be preserved for future generations. Natural resources are the air we breathe, the water we drink, the soil that supports plant and animal life and the minerals under the ground. All of these nonhuman-made resources are vital to the well being of nations. Ecologists think of natural resources as *renewable* and *nonrenewable*. Forests, grasslands and the wild species that inhabit them can be restored after their numbers have been somewhat depleted. These are the renewable resources. Minerals, including the fossil fuels of oil and coal, cannot be restored once used up and therefore are classified as nonrenewable. Desirable conservation practices must be the concern of every human. Understanding principles of ecology and of conservation is paramount to the making of wise decisions.

SOIL CONSERVATION

The soil covering the earth is really a very thin layer measuring about 0.9 meter in depth. Soil is not made quickly. It takes thousands of years, much erosion and weathering of rock, and the gradual addition of plant and animal remains to make a fertile top soil. Top soil can be quite fragile and easily destroyed by *sheet erosion*, washed away from sloping hills by rain. Soil may also become exhausted or *leached*, having been over-cultivated so that the mineral components are lost.

Conservation measures protect top soil. Erosion can be prevented by the wise use of ground cover, rooted plants that hold the soil in place. *Terracing*, the planting of trees, bushes and low-growing plants in steps cut into the hillsides prevents loss of soil by sheet erosion. *Contour plowing*, the circular plowing of hillsides, is another method of preventing the erosion of soil. The leaching of minerals and nitrates from the soil is prevented by *strip cropping* (planting corn or cotton with alternate rows of legumes) and *crop rotation*, planting a field on alternate years.

CONSERVATION OF WATER

Water is rendered useless for drinking, for bathing, for irrigation and as a habitat for fish when it is polluted by the chemical wastes from industry and by human sewage. Sewage treatment plants clean up sewage before it is dumped into waterways. Special treatment must be given to chemical wastes to detoxify them before disposal.

Water is a renewable resource. However, people on the earth are using more water than even before in industry, refrigeration, agriculture and the like. Humans are dependent on rainfall to maintain an adequate *water table* (level of groundwater) and to replenish water stores in reservoirs. The wasting of water through careless use can have serious consequences for human life.

CONSERVATION OF FORESTS

Forests are important in the conserving of water, soil and habitats for wild life. Forests also supply us with lumber and natural resins that are used in the paint industry and for other industrial processes. Forests are destroyed by over-lumbering through which more trees are cut down than can be replaced by natural processes. Trees grow slowly, requiring about thirty or forty years to reach maturity. We need wood to make paper and to build houses, telephone poles and fences. However, if forests are lumbered imprudently, trees will not last. Most of the virgin forests in the United States have been destroyed, and now in their places are second growth forests.

Chronology of Famous Names in Biology

1840	**Justus von Liebig** (German)—developed the ecological concept of the Law of the Minimum in which he showed the limiting effect on growth of environmental factors.
1869	**Ernst Haeckel** (German)—coined the word "ecology" as the science that studies environment.
1920	**W. W. Garner** and **H. A. Allard** (American)—discovered that daylength affects the flowering of the Biloxi soybean and also the flowering of a mutant form of tobacco called "Maryland Mammoth".
1938	**K. C. Hamner** and **J. Bonner** (American)—discovered how the cockleburr responds to the photoperiod.
1939	**George Washington Carver** (American)—discovered that legumes such as peanuts returned nitrates and minerals to the soil. He developed 300 new industrial uses for peanuts and 118 byproducts from sweet potatoes.
1942	**Elton Charles** (American)—made a detailed study of irruption and death of lemmings.
1948	**Ralph Buchsbaum** (American)—wrote a very important work on basic ecology.

1952   **Rachel Carson** (American)—wrote vivid descriptions of the oceans as ecosystems and about the organisms that live therein.

1955   **F. W. Went** (American)—a modern botanist, who has elucidated the various roles of plants in ecosystems.

1955   **Herbert Zim** (American)—presented important information about the living organisms that inhabit seashores.

1957   **G. S. Avery** (American)—elucidated reasons for the death of oak trees in U.S. cities.

1958   **Eugene Odum** (American)—a modern researcher who has been a motivator in the field of ecology.

1959   **W. H. Amos** (American)—researched the living organisms of sand dunes.

1960   **Howard T. Odum** (American)—developed techniques for measuring the energy flow in forests and marine environments.

1960   **Marston Bates** (American)—a leading authority on the ecology of world biomes.

1964   **Rene Catala** (New Caledonian)—an authority on fluorescent corals.

1967   **George M. Woodwell** (American)—leading authority on the effects of radiation and toxic substances on ecosystems.

Words for Study

abiotic	ecology	photoperiodism
autotroph	ecosystem	primary consumer
biosphere	erosion	producer
biotic	food chain	scavenger
carnivore	food web	secondary consumer
climax	food pyramid	succession
community	herbivore	taiga
conservation	heterotroph	trophic level
crop rotation	limiting factor	tolerance
cycling	natural resources	tundra
deciduous	niche	
decomposers	nitrogen fixation	

Questions for Review

PART A.   Completion. Write in the word that correctly completes each statement.

1. The living factors in the environment are known as ..1.. factors.
2. The living community and the nonliving environment are known collectively as the ..2..
3. Without the work of ..3.. the remains of dead organisms would pile up.
4. A nuthatch is a species of ..4..
5. The number of species that can successfully occupy a niche is ..5..
6. The flow of energy through an ecosystem can be studied by identifying simple ..6.. (two words).
7. An herbivore is also classified as a ..7.. consumer.
8. A great deal of water is lost from plants by an evaporation process known as ..8..
9. About 80 percent of the volume of air is the elemental gas ..9..
10. Among the limiting factors in the environment are light, temperature and ..10..
11. Phytochrome is a plant hormone that controls ..11..
12. The seeds of most deciduous plants must go through a period of ..12.. before they can germinate.
13. The compound that all living things need for the dissolving of chemical substances is ..13..
14. The orderly change of a biotic community is known as ..14..
15. A geographical area having one type of climate is a ..15..

PART B.   Multiple Choice. Circle the letter of the item that correctly completes each statement.

1. The nonliving factors in an ecosystem are referred to as
   (a) adiabatic   (c) abiotic
   (b) abomasum   (d) abscissant

2. What percent of the air is made up of oxygen?
   (a) 21   (c) 78
   (b) 50   (d) 90

3. The habitat of phytoplankton is
   (a) deciduous forest   (c) Nairobi desert
   (b) ocean surface   (d) Canadian taiga

4. An ecological niche is a (an)
   (a) organism   (c) habitat
   (b) geographical region (d) feeding pattern

5. The lives of organisms in an ecosystem depend upon the
   (a) amount of light in the ecosystem
   (b) quality of environment
   (c) availability of pH in the ecosystem
   (d) flow of energy through the ecosystem

6. In a food chain, lettuce is best classified as a
   (a) producer   (c) vegetable
   (b) deciduous plant   (d) detritus

7. The arrow in a food web diagram stands for
   (a) alternative energy pathway (c) is eaten by
   (b) increased energy   (d) eats up

8. In a food web, hawks are best classified as
   (a) scavengers   (c) prey
   (b) predators   (d) primary consumers

9. In plant cells, phytochrome is associated with
   (a) nuclei   (c) lysosomes
   (b) cytoplasm   (d) plasma membranes

10. The washing away of top soil from the slopes of hills is known as
    (a) weathering   (c) irrigation
    (b) erosion   (d) leaching

11. An example of an adaptation that permits the survival of some plant species in the tropical rain forest is
    (a) degenerate stomates (c) aerial roots
    (b) needlelike leaves   (d) cutinized stems

12. Chemical changes on the surfaces of rock may be brought about by
    (a) lichen colonies   (c) tree roots
    (b) heavy rainfall   (d) germinating bryophytes

13. Musk ox, caribou and foxes are part of the mammal population of the
    (a) temperate forest   (c) grasslands
    (b) taiga   (d) tundra

14. In desert biomes, the average rainfall is no greater than
    (a) 20 cm   (c) 15 cm
    (b) 6.5 cm   (d) 0 cm

15. Examples of primary consumers in the marine biome are
    (a) halibut and cod   (c) sardines and anchovies
    (b) rays and haddock   (d) cod and haddock

PART C.   **Modified True-False.** If a statement is true, write "true" for your answer. If a statement is incorrect, change the underlined word to one that will make the statement true.

1. The cultivation of a field in alternate years is known as crop rotation.

2. Oil, coal and natural gas are renewable resources.

3. Straight row plowing prevents the erosion of hillsides.

4. All of the plant and animal species interacting in a given environment are known as a population.

5. Tertiary consumers are those that feed on smaller <u>herbivores</u>.

6. Vultures are best classified as <u>predators</u>.

7. In ecosystems, fungi and bacteria serve in the roles of <u>scavengers</u>.

8. Competition results when two species share the same <u>habitat</u>.

9. Once energy is used, it is converted into <u>fat</u>.

10. A process in which energy is released is <u>photosynthesis</u>.

11. Most of the energy in a food chain is concentrated at the <u>carnivore</u> level.

12. The alternative pathways of energy flow in an ecosystem are best shown in a food <u>pyramid</u>.

13. The relationships between predators and prey help to <u>disrupt</u> an ecological community.

14. The source of hydrogen utilized by green plants for photosynthesis is <u>sugar</u>.

15. A <u>changing</u> community is known as a climax community.

Answers to Questions for Review

PART A

1. biotic
2. ecosystem
3. decomposers
4. bird
5. one
6. food chains
7. primary
8. transpiration
9. nitrogen
10. rainfall
11. flowering
12. cold
13. water
14. succession
15. biome

PART B

1. c	6. a	11. c
2. a	7. c	12. a
3. b	8. b	13. d
4. d	9. d	14. b
5. d	10. b	15. c

PART C

1. true
2. nonrenewable
3. Contour
4. community
5. carnivores
6. scavengers
7. decomposers
8. niche
9. heat
10. respiration
11. autotroph
12. web
13. stabilize
14. water
15. stable

INDEX